MYSTERIOUS NATURE OF TIME

A JOURNEY THROUGH THE FABRIC OF REALITY

PRADYUMN

ISBN
Paperback 979-8-89544-294-4
Hardcase 979-8-89544-814-4

CONTENTS

FOREWORD

Time, in its relentless march, has long captivated the minds of scientists and philosophers alike. As we move through our daily routines, governed by the steady ticking of the clock, we rarely stop to consider the profound mysteries and implications of this ever-present dimension. The book you now hold, "Mysterious Nature of Time: A Journey Through the Fabric of Reality," invites you to embark on a profound exploration of time, guided by the insightful mind of Pradyumn.

In this work, Pradyumn skillfully guides us from the familiar rhythms of our everyday lives to the awe-inspiring theories of Einstein's relativity and the enigmatic realms of quantum mechanics. Each chapter showcases his remarkable ability to distill complex ideas into accessible and engaging narratives, making intricate scientific concepts approachable for readers of all backgrounds. From the fundamental aspects of time and its measurement to philosophical inquiries into its nature and the tantalizing possibilities of time travel, Pradyumn navigates these topics with clarity and depth.

Having already established himself as a formidable author with previous works like "Journey through the Dark Monster" and "The World of Particles," Pradyumn continues to inspire and enlighten with this latest offering.

His extensive research on cutting-edge topics such as thermoelectric materials, supercapacitor electrodes, bandgap engineering, spintronics, and doping effects using DFT, along with his academic presentations and publications, underscores the depth of his expertise.

"The Mysterious Nature of Time" is more than just a scientific treatise; it is a journey that challenges our perceptions and broadens our understanding of the universe. Pradyumn's narrative is not only informative but also deeply inspiring, urging readers to embrace the wonders and complexities of time.

As you delve into the pages of this book, prepare to be both enlightened and inspired. Pradyumn's exploration of time will undoubtedly leave you with a renewed sense of wonder and curiosity about the very fabric of reality. Fasten your seatbelts and enjoy this captivating odyssey through the unfathomable mysteries of time.

Dr. Archana Sharma
Pravasi Bhartiya Awardee 2023 by
Government of India
Senior Scientist at CERN, Geneva

PREFACE

Time is an ever-present, relentless force that governs our existence. It's a concept we're intimately familiar with, as we constantly measure it in seconds, minutes, and hours. We live our lives by the ticking of the clock, scheduling our activities, and planning for the future. Yet, despite our daily interactions with time, it remains one of the most enigmatic and elusive concepts in the universe.

In this book, "Mysterious Nature of Time: A Journey Through the Fabric of Reality," we embark on an exploration of time that takes us from the familiar ticking of our wristwatches to the mind-bending concepts of Einstein's theory of relativity and the quantum realm. We'll delve into the profound philosophical questions surrounding time, contemplate the possibility of time travel, and consider the cosmic implications of this fundamental dimension.

The journey through time is not just a scientific endeavor; it's a philosophical and existential one. We'll traverse the terrain of time's mysterious nature, challenging the boundaries of our understanding and pushing the limits of our imaginations.

Our quest begins with the foundational aspects of time in Chapter 1, where we introduce the very basics of this dimension and explore the different ways we measure it. We discuss the arrow of time, a concept that underpins much of what follows, leading us to the heart of the timekeeping paradox.

In Chapter 2, we venture into the realm of relativity, where Albert Einstein's groundbreaking theories forever changed our perception of time. Special relativity teaches us that time is relative, and the faster you move, the slower time appears to pass. General relativity deepens our understanding by revealing the profound influence of gravity on time.

As we journey further, we'll enter the mysterious world of quantum mechanics in Chapters 3 and 4, where time takes on an entirely different character. Here, we explore the uncertainty principle, entanglement, and their implications for our understanding of time.

Our exploration continues with the flow of time in Chapters 5 and 6, addressing the profound philosophical questions about time's direction and the tantalizing possibility of time travel.

In Chapters 7 and 8, we venture into the realms of philosophy and cosmology, seeking answers to questions about the nature of time and its relationship with the cosmos.

In the final stretch, we contemplate the future of time in Chapter 9, examining its role in technology and the potential impact of emerging technologies on our perception of time. In Chapter 10, we confront the unresolved questions about time, embracing the mysteries that remain.

As we embark on this journey through the fabric of reality, my hope is that you'll not only gain a deeper understanding of time but also find inspiration in the wonders and complexities that this concept holds. So, dear reader, fasten your seatbelts as we embark on this captivating odyssey through the unfathomable mysteries of time.

ABOUT THE AUTHOR

Fascinated by the beauty of Physics in class 10th, Pradyumn started exploring its wonders and sharing them with the world through his blog named 'Physics Mindboggler (**PM**)'. The humble blog transformed into an impact-making Social Venture in his 3rd year of B. Tech. at Bennett University. **PM** today strives to inspire young minds towards Science and to connect Research with Entrepreneurship. Recently **PM** was honored with the 'Best Startup Award' by the World Association for SMEs.

Pradyumn completed his graduation in B. Tech. in Engineering Physics from Bennett University (Times of India Group) with the 'Chancellor's Medal for Innovation and Entrepreneurship' received at the hands of the Bollywood superstar Ranveer Singh.

Pradyumn the author of two books titled 'Journey through the Dark Monster' and 'The World of Particles'. Published in his teens, the former is a primer on general relativity and black holes and the latter unravels the recipe to build our universe. During his Engineering days, he also published various research papers in international conferences and journals.

He has done research work on thermoelectric materials, supercapacitor electrode, bandgap engineering, spintronics, and doping effect using DFT. He presented his first research paper in his 1st year of B. Tech. in WEENTECH International conference at Heriot-Watt University, United Kingdom. His recent paper was published as a book chapter in a Springer publication. With this and upcoming books and research works, Pradyumn is on his way to achieving his dreams.

Fully aware that all work and no play make Jack a dull boy, Pradyumn is also a Second Dan Black Belt in Karate Budokan International.

ACKNOWLEDGMENT

After positive response from my first two books, many teachers and young minds encouraged me work on my third book. I was always fascinated with the role time plays in macroscopic and microscopic realms. Hence, I started exploring its mysterious nature. However, the journey to explore the mysterious nature of time and present them in this book, would not have been possible without the unwavering support and encouragement of numerous other individuals. I would like to acknowledge all of them for their support and blessings.

The genesis of this book can clearly be traced back to my teachers at the St. Pius X High School and Vani Vidyalaya Jr. College, who inspired me to explore the realm of science and technology. The person who ignited the passion in me to raise fundamental questions and to seek answers to them through observation and experimentation was Prof. Shiv Yadav, my Physics teacher at Akhil Tutorials. In fact, this book is a result of the passion ignited by him.

All my teachers since then, right up to my alma matter teachers, HoD, Dean, and Vice-Chancellor at Bennett University, have only fuelled that fire. It would be as futile to even start thanking all these gurus, as it would be to thank my beloved parents and family. No words of gratitude can do justice to their contribution in making me who I am.

I express special gratitude towards my mentors at Bennett Hatchery Dr. Vinod Shastri and Mr. Manish Mathur. The book would definitely have been written, but without their constant encouragement and support, it probably would not have been published so soon.

I thank Jasreet Kaur Bachhal of Notionpress for her guidance and support in publishing this book and taking it to discerning readers like you.

Last, but not the least, I express special gratitude towards all my well-wishers who financially supported the publication of this book through a fundraiser created on Milaap, the largest crowdfunding platform in India. I am equally thankful to numerous others, who insisted on remaining anonymous.

– Pradyumn

List of Financial Supporters

Akanksha Mahadik	Kamalakar Mhatre	Sahil Sawant
Amit Raje	Nitant Ashok Tambade	Sanjay D. Bangar
Anmol Goyal	Pankaj Kumar	Sneha Kedar
Atharv Bhange	Parag Gore	Sujaya Rao
Avinash Mohite	Pradyumn Jain	Sunanda Mane
B. S. Mane	Prashant S. Kadam	Sunil Vithal Parab
Diksha Talgaonkar	Dr. Pratik Das	Utkarsha Chavan
Dilip Impal	Pushpa Jhadhao	Valay Kumar Jain
Dr. Manthan Pokharkar	Rahul Nambiar	Dr. Vinod Shastri
Ishanvi Mohite	Revati Kadam	
Jayashree Subhash Parab	Rinkal Karia	

Chapter 1

THE TICKING CLOCK - UNDERSTANDING THE BASICS

We begin our journey through the mysterious realm of time by exploring its most basic and fundamental aspects. Time is a concept we experience every day, yet it's one of the most profound and intricate ideas in the universe. In this chapter, we'll delve into the essential components of time, the history of timekeeping, and the fascinating insights that have shaped our understanding of this enigmatic dimension.

A Brief History of Time

Let's step back in time, not too far, to a place where the concept of time first took root. Our ancestors, thousands of years ago, had an intuitive grasp of the passage of time. They observed the regular cycle of day and night, the changing seasons, and the phases of the moon. These natural phenomena became their first timekeepers.

Figure 1.1: A sundial, one of the earliest timekeeping devices, with the sun casting a shadow to indicate the time of day

The Sundial: Around 1500 BC, the ancient Egyptians and Babylonians created the first sundials. These devices used the movement of the sun's shadow to tell the time. Imagine a stick or obelisk casting a shadow on a marked dial as the sun traveled across the sky. By reading where the shadow fell, people could estimate the time. Sundials were simple, effective, and crucial for organizing daily life.

The Quest for Precision

As societies evolved, so did the need for more precise timekeeping. Sundials had their limitations, as they were dependent on daylight and didn't work at night or on cloudy days. To address these issues, different cultures developed their own timekeeping methods.

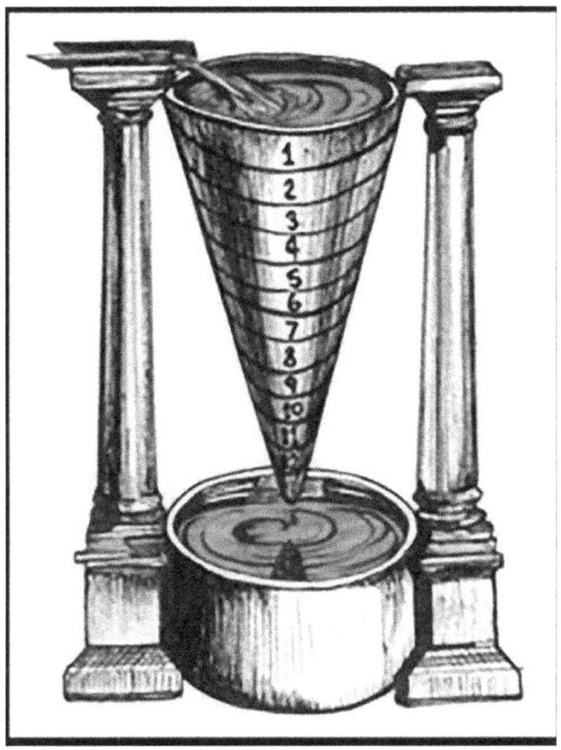

***Figure 2: An ancient water clock, or clepsydra, with water flowing
from one container to another to measure time***

Water Clocks (Clepsydra): The ancient Greeks and Chinese, among others, developed water clocks. These clocks used the steady flow of water from one container to another to measure time. The level of water in the receiving container indicated the time elapsed. While an improvement over sundials, water clocks had their own challenges, such as variations in water flow due to temperature changes.

Mechanical Marvels

It wasn't until the Middle Ages that mechanical clocks made their appearance, marking a significant leap in timekeeping technology. One of the most famous early examples is the astronomical clock in the Strasbourg Cathedral, constructed in 1354.

Figure 1.3: The astronomical clock in Strasbourg Cathedral. Credit: DAVID ILIFF, CC BY-SA 3.0

These mechanical clocks incorporated intricate gear systems and were often accompanied by elaborate astronomical displays. They were housed in towers and provided both practical timekeeping and a beautiful showcase of the craftsmanship of the time.

Pendulum Power

The 17th century brought a revolution in timekeeping with the invention of the pendulum clock by Dutch scientist Christiaan Huygens. This remarkable innovation significantly improved the accuracy of clocks. The pendulum's regular back-and-forth motion allowed for more precise time measurement.

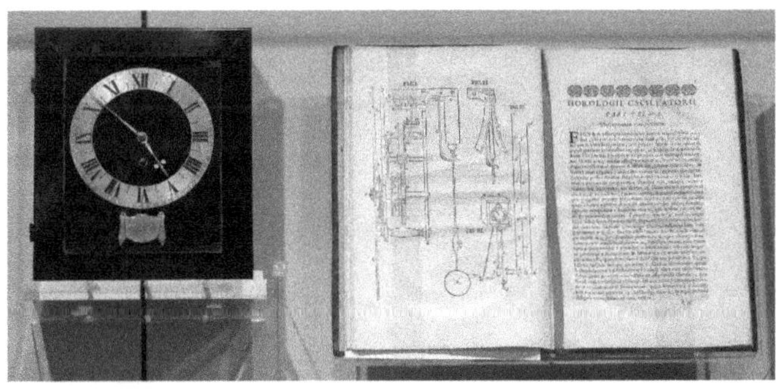

Figure 1.4: Spring-driven pendulum clock, designed by Huygens and built by Salomon Coster (1657) with a copy of the Horologium Oscillatorium (1673) at Museum Boerhaave, Leiden. Credit: Rob Koopman, CC BY-SA 1.0

The discovery of the pendulum's isochronous properties, meaning it swings with a constant period, led to a dramatic improvement in timekeeping precision. These pendulum clocks became the standard for measuring time in homes and public places.

The Birth of Standard Time

As humanity advanced, the need for synchronized time across regions and countries became increasingly important. Before the introduction of standard time, each town and city would set their local time based on the position of the sun. This system was practical when people didn't travel far from home, but with the advent of the railway in the 19th century, it became evident that a more standardized approach was needed.

Figure 1.5: A vintage railway station clock. Credit: K. Krallis (SV1XV), CC BY-SA 3.0

Sir Sandford Fleming, a Canadian railway planner, proposed the division of the world into 24 time zones, each one-hour apart, which would later become the basis for Coordinated Universal Time (UTC). On November 18, 1883, the concept of standard time was implemented, forever changing the way we think about time.

The Atomic Revolution

While mechanical clocks were incredibly accurate for their time, they still had limitations. The quest for even greater precision led to a remarkable breakthrough in the mid-20th century: the development of atomic clocks.

Figure 1.6: UK National Physical Laboratory cesium fountain clock.
Credit: National Physical Laboratory, United Kingdom

Atomic clocks use the vibrations of atoms, typically cesium or rubidium, to keep time. These vibrations are incredibly stable and serve as the foundation for the definition of the second, the base unit of time in the International System of Units (SI). The precision of atomic clocks is staggering. They can keep time to within a few billionths of a second per day.

The Time We Live By

Today, atomic clocks are at the heart of modern timekeeping. The time you see on your smartphone, computer, or any other electronic device is most likely based on the highly accurate measurements of atomic clocks. We've come a long way from the sundials and water clocks of ancient times, and the progression of timekeeping technology has been nothing short of astounding.

But what is time, fundamentally? It's not just the numbers on a clock or the cycles of celestial bodies. Time is a dimension that pervades the entire universe, and its nature has been a subject of deep contemplation and scientific inquiry for centuries.

The Nature of Time

At its core, time is a measure of change. It's a way to order events and compare their durations. We can define time as the progression from the past, through the present, and into the future. It's a fundamental aspect of our experience, and it shapes our daily lives.

Imagine this: you wake up in the morning, and the clock reads 7:00 AM. Throughout the day, you mark the passage of time as you go about your activities. You attend meetings, have lunch with a friend, and watch the sunset. Each of these events occurs in a specific sequence, one after the other, and you use time to order them.

Time is what allows us to make sense of our world. It's the rhythm to which we dance through life, a beat that never stops. It's the canvas upon which we paint the story of our existence.

The Arrow of Time

As we ponder the nature of time, one question naturally arises: Does time have a direction? In other words, is there an arrow of time that points from the past to the future, and if so, why?

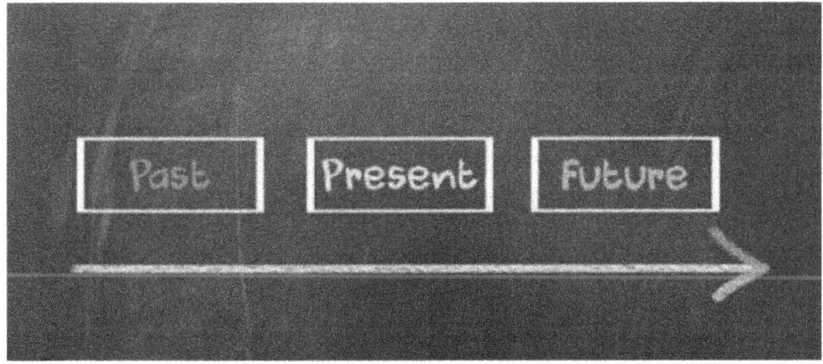

Figure 1.7: An illustration of the arrow of time, depicting the progression from past to present to future

The arrow of time is a fascinating and deeply philosophical aspect of time. It's what allows us to remember the past, experience the present, and anticipate the future. Without this arrow, our reality would be a jumbled mess of events with no order or coherence.

However, understanding the nature of this arrow is a profound challenge. One of the most compelling explanations comes from the realm of thermodynamics.

The Complexity of Time

Time is an extraordinary concept, and even in this introductory chapter, we've touched on some of its most profound and complex aspects. From the simple sundial to atomic clocks and the arrow of time, time spans the spectrum of human understanding, from the intuitive to the deeply mysterious.

In the pages that follow, we'll continue our exploration of time, delving deeper into the intricate and often mind-bending aspects of this fundamental dimension. We'll uncover the mysteries of time's relativity, the quantum world, and the possibilities of time travel. We'll also journey through the realms of philosophy and cosmology, contemplating the cosmic implications of time.

As we navigate this incredible voyage through time, my hope is that you'll not only gain a deeper understanding of the concept but also find inspiration in the profound mysteries that surround it. Time is not just a tool we use to schedule our daily lives; it's a window into the very fabric of reality itself. It's the unifying thread that weaves the past, present, and future into the tapestry of existence.

Chapter 2

THE RELATIVITY OF TIME - EINSTEIN'S REVOLUTION

In the previous chapter, we explored the history of timekeeping, from ancient sundials to atomic clocks, and the profound concept of the arrow of time. Now, we venture into one of the most revolutionary and mind-expanding ideas in the realm of physics: Einstein's theory of relativity. This theory not only challenged our understanding of time and space but also reshaped the very fabric of the cosmos.

The Need for a New Theory

In the late 19th and early 20th centuries, scientists were grappling with a conundrum. Two of the most successful theories in the history of science—Newton's laws of motion and the theory of electromagnetism—appeared to be in conflict. These theories worked exceptionally well in their respective domains, but when applied to phenomena involving very high speeds or strong gravitational forces, they produced conflicting results.

Figure 2.1: A portrait of Albert Einstein

This conflict between classical physics and the observed behavior of the universe presented a fundamental challenge. Scientists needed a new framework that could reconcile these inconsistencies and provide a more accurate description of reality. Enter Albert Einstein.

Special Relativity: A New Paradigm

In 1905, Albert Einstein introduced his theory of special relativity, which revolutionized our understanding of time, space, and motion. This theory laid the foundation for the more comprehensive theory of general relativity, but we'll start with the special version to grasp the essential concepts.

Einstein's theory of special relativity introduced two groundbreaking postulates:

1. The laws of physics are the same for all observers, regardless of their relative motion.

2. The speed of light in a vacuum (denoted as "c") is the same for all observers, regardless of their motion or the motion of the source of light.

These two postulates might seem simple, but their implications are profound. They challenged the long-held Newtonian idea that time and space were absolute, fixed entities.

Time Dilation

One of the most mind-bending consequences of special relativity is time dilation. It's the idea that time is not a universal constant but rather depends on the relative motion of observers. To understand this concept, let's explore a famous thought experiment known as the "twin paradox."

Figure 2.2: A thought experiment illustrating the twin paradox

Imagine two identical twins, Alice and Bob. Alice stays on Earth, while Bob boards a spaceship and embarks on a journey traveling at a significant fraction of the speed of light. When Bob returns to Earth, he finds that he has aged less than Alice. In other words, time has passed more slowly for Bob.

This is not science fiction but a real phenomenon predicted by special relativity. According to the theory, the faster an object moves relative to an observer, the more time slows down for that object. In this case, Bob's high-speed journey caused time to dilate for him, resulting in less aging compared to Alice on Earth.

The twin paradox highlights the counterintuitive nature of time dilation. From Bob's perspective on the spaceship, it's Alice who is moving at a high speed, and therefore, she should experience time dilation. However, both perspectives are valid within the framework of special relativity. This demonstrates the relative nature of time and the critical role of the speed of light.

Spacetime: A Unified Fabric

Special relativity not only introduced time dilation but also reshaped our understanding of the relationship between time and space. Einstein's theory merged them into a unified fabric known as "spacetime." In this four-dimensional framework, an event is specified not just by its location in space but also by its position in time.

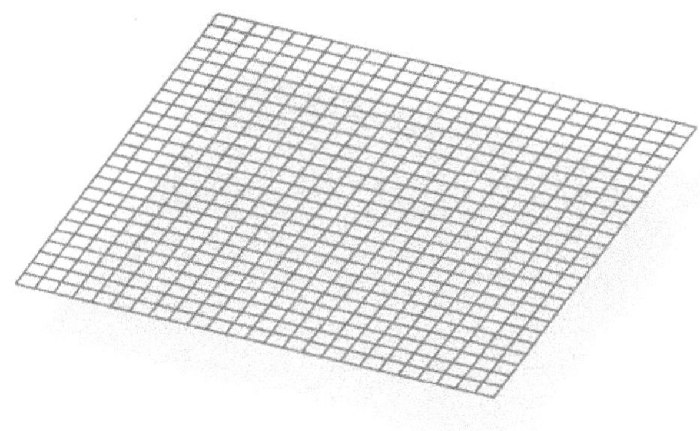

Figure 2.3: A visual representation of spacetime, where time and space are intertwined, creating a unified fabric

Spacetime provides a new way to conceptualize reality, where time and space are intertwined and inseparable. The path of an object through spacetime is called its worldline, and it's this worldline that describes an object's entire history, from its past to its future.

Imagine you toss a ball into the air. Its worldline traces the path it follows through spacetime, curving downward due to gravity and upward as it slows and reverses direction. This curvature represents the influence of spacetime on the ball's motion, as described by general relativity, which we'll explore in depth in the next section.

General Relativity: The Curvature of Spacetime

While special relativity dealt with relative motion and the speed of light, it was Albert Einstein's theory of general relativity that fundamentally altered our understanding of gravity and the structure of the universe. General relativity describes gravity not as a force between objects, as described by Newton, but as the curvature of spacetime caused by massive objects.

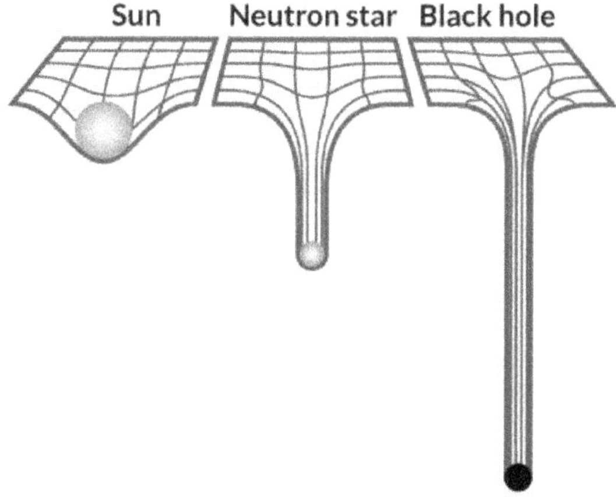

Figure 2.4: An illustration of how massive objects warp the fabric of spacetime. Credit: James Provost, sciencenews.org

Einstein's theory of general relativity was published in 1915 and was a culmination of a decade of intense scientific exploration. It was tested and confirmed during a solar eclipse in 1919 when astronomers observed the bending of starlight as it passed near the Sun. This observation provided strong evidence in favor of the theory and made Einstein a household name.

General relativity introduces the concept of a gravitational field as a result of massive objects bending the fabric of spacetime around them. When smaller objects, like planets, move within this curved spacetime, they follow curved paths, which we perceive as gravitational attraction.

The Rubber Sheet Analogy

To understand the concept of spacetime curvature, think of it as a rubber sheet. When a massive object, like a bowling ball, is placed on the sheet, it creates a depression. Now, imagine rolling a smaller object, like a marble, near the bowling ball. The marble will follow a curved path around the depression, much like a planet orbiting a star.

Figure 2.5: An analogy of spacetime curvature using a rubber sheet and a heavy ball creating a depression. Credit: Dan Burns

This analogy simplifies the idea of how gravity works in the context of general relativity. It's not a mysterious force pulling objects toward each other; instead, massive objects cause spacetime to curve, and other objects follow these curved paths. This concept beautifully explains the orbits of planets, the bending of light by gravity, and other gravitational phenomena.

However, the rubber sheet analogy, while is a helpful visualization for gravity in general relativity, it has limitations, such as oversimplifying the multidimensional nature of spacetime, implying a force rather than curvature, presenting a static deformation, lacking a physical explanation, and omitting the time dimension. It serves as an introductory concept but doesn't fully capture the complexity of gravitational interactions.

Gravitational Time Dilation

In addition to curvature, general relativity predicts another form of time dilation: gravitational time dilation. This effect occurs because the strength of gravity varies with an object's distance from a massive body. The closer you are to a massive object, the stronger the gravitational field, and the slower time passes for you.

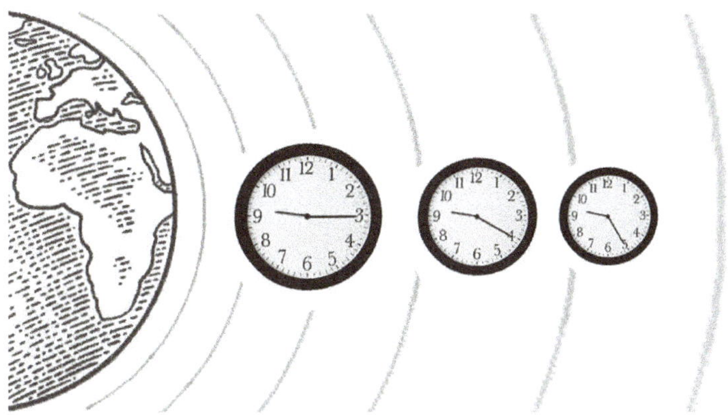

Figure 2.6: An illustration showing how gravitational time dilation causes time to pass more slowly near a massive object, like a planet

One of the most famous demonstrations of gravitational time dilation occurs with the Global Positioning System (GPS). The satellites in the GPS system are in orbit around Earth, and because they are farther from the planet's gravitational center, time passes more quickly for them than it does on the surface of the Earth. Without correcting for this effect, the GPS system's accuracy would suffer, resulting in location errors of several meters.

The fact that we must account for gravitational time dilation in everyday technologies showcases how central these concepts are to our modern understanding of time and space.

Black Holes: Time's Ultimate Twist

The most extreme consequences of general relativity occur in the vicinity of black holes. A black hole is a region in spacetime where gravity is so intense that nothing, not even light, can escape its grasp. The boundary around a black hole is called the event horizon.

Figure 2.7: An artistic representation of a black hole. Credit: ESO, ESA/ Hubble, M. Kornmesser/N. Bartmann

Black holes are often portrayed as cosmic vacuum cleaners that indiscriminately swallow everything in their path, including time itself. The closer you get to a black hole, the stronger the gravitational field, and the more time slows down. If you were to watch someone approaching a black hole from a safe distance, you'd see them move more and more slowly as they approached the event horizon. To the person nearing the black hole, time would appear to pass as usual, but to distant observers, it would seem to stand still.

This phenomenon is a direct consequence of the extreme spacetime curvature near a black hole and is known as "gravitational time dilation." It's one of the most mind-bending aspects of general relativity and underscores the profound impact that gravity can have on the passage of time.

The Relativity of Reality

Einstein's theories of relativity transformed our understanding of time and space, taking us from the idea of absolute time and space to a universe where they are inextricably linked. Special relativity taught us that time is relative, and its passage depends on an observer's motion. General relativity revealed that gravity is not a force but the result of spacetime curvature caused by massive objects.

These ideas may seem far removed from our everyday experience, but they have practical consequences and have reshaped our modern world. Our understanding of GPS systems, the functioning of satellites, and even our understanding of the cosmos is deeply intertwined with the principles of relativity.

As we journey further into the exploration of time in the chapters to come, we'll encounter even more fascinating and mind-expanding concepts. We'll delve into the quantum realm of time, ponder the mysteries of time travel, and contemplate the cosmic implications of this profound dimension.

The insights of Einstein's relativity theories remind us that reality is often more nuanced and complex than our everyday experiences suggest. They challenge us to think beyond our intuitions and to explore the profound mysteries of time and space. In doing so, we not only gain a deeper understanding of the universe but also find inspiration in the boundless horizons of scientific discovery.

Chapter 3

TIME IN THE QUANTUM WORLD

In the previous chapters, we explored the history of timekeeping and delved into the revolutionary theories of special and general relativity, which reshaped our understanding of time, space, and gravity. Now, our journey through the complexities of time leads us into the mind-bending realm of quantum mechanics, where the conventional rules of time take on a new and enigmatic form. In this chapter, we'll explore the quantum nature of time and how it challenges our everyday intuitions.

The Quantum Revolution

The birth of quantum mechanics in the early 20th century marked a profound shift in our understanding of the physical world. Scientists like Max Planck, Albert Einstein, Niels Bohr, and Erwin Schrödinger, among others, ushered in a new era of physics, where the classical certainties of Newton's laws and electromagnetism gave way to the probabilistic and uncertain nature of the quantum realm.

Figure 3.1: A photograph of the key figures in the development of quantum mechanics. On the top (from left to right) Neils Bohr, Albert Einstein, and Max Planck. On the bottom (from left to right) Wolfgang Pauli, Werner Heisenberg, and Erwin Schrödinger

At the heart of quantum mechanics is the idea that at the smallest scales of the universe, matter and energy exhibit wave-particle duality, where they can behave both as particles and waves. This duality introduces a level of uncertainty that challenges our everyday experience of the world.

The Quantum of Time

In the quantum world, time is no exception to the strange and mysterious behavior exhibited by particles and energy. One of the foundational principles of quantum mechanics, Heisenberg's Uncertainty Principle, tells us that the more precisely we know a particle's position, the less precisely we can know its momentum, and vice versa.

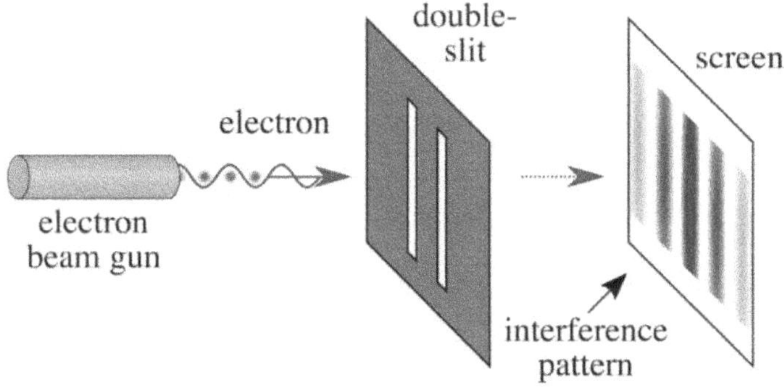

Figure 3.2: Double-slit experiment demonstrating the Heisenberg Uncertainty Principle, where precise measurement of position increases uncertainty in momentum. Credit: NekoJa, Johannes Kalliauer, CC BY-SA 4.0

This principle extends to time as well. It implies that if we know the exact time of an event, we cannot precisely determine the energy associated with that event. Conversely, if we have a precise measurement of energy, the exact time of that event becomes uncertain. This inherent uncertainty in time and energy measurements forms the basis of the quantum of time.

The concept of the quantum of time is intimately linked to Planck's constant, denoted as "h." Planck's constant is a fundamental constant of nature and plays a central role in quantum mechanics. It sets the scale for the smallest units of action and energy in the universe.

Figure 3.3: A portrait of Max Planck

Planck's constant is incredibly small, approximately 6.626×10^{-34} joule-seconds, which means that in the macroscopic world we inhabit, the effects of this quantum of action are negligible. However, when we venture into the quantum realm, where particles and energies are extremely tiny, the effects become significant.

The quantum of time, also known as the Planck time (t_p), is a fundamental time scale defined using Planck's constant and other constants of nature. It's the time it takes for a photon (a particle of light) to travel a distance equal to the Planck length (the smallest length scale according to quantum theory), at the speed of light (c).

The Planck time is incredibly brief, approximately 5.391 x 10^{-44} seconds. To put this into perspective, it's a time scale so short that it's far beyond our ability to measure with current technology. It's more than 20 orders of magnitude shorter than the shortest time intervals we can currently measure in particle physics experiments.

The Quantum Realm: Schrödinger's Cat

To illustrate the curious nature of quantum uncertainty, let's explore a famous thought experiment devised by Erwin Schrödinger, known as "Schrödinger's Cat."

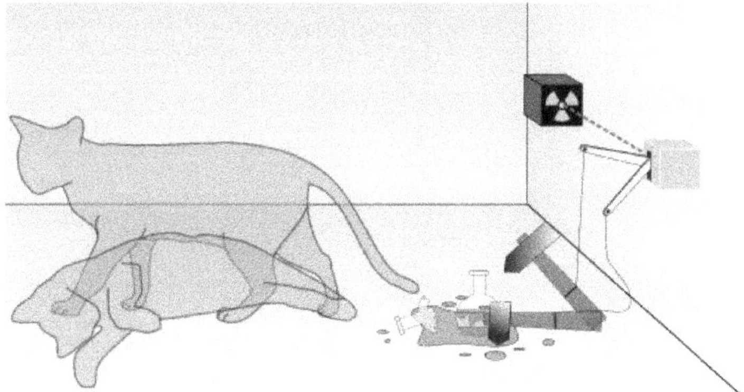

Figure 3.4: An illustration of Schrödinger's cat experiment. Credit: Dhatfield, CC BY-SA 3.0

In this scenario, imagine a sealed box containing a cat, a radioactive atom, and a vial of poison. The radioactive atom has a 50% chance of decaying within a specific time frame. If it decays, it triggers the release of the poison and the cat's demise; if it doesn't decay, the cat survives.

According to quantum mechanics, before we open the box to observe the cat, the system is described by a quantum superposition—a combination of both "live" and "dead" states. It's as if the cat is in a state of being both alive and dead at the same time. Only when we open the box and make an observation does the superposition collapse, and the cat is either alive or dead.

This concept challenges our classical intuitions, as it suggests that until we make a measurement, the outcome is uncertain and exists in a superposition of possibilities. It's as though time and reality in the quantum realm exist in multiple states simultaneously until we interact with the system.

Time in Quantum Mechanics

In quantum mechanics, the concept of time takes on a unique character. Instead of the classical notion of time as a continuous and absolute dimension, time in quantum mechanics is treated as an operator—an observable quantity that can be measured.

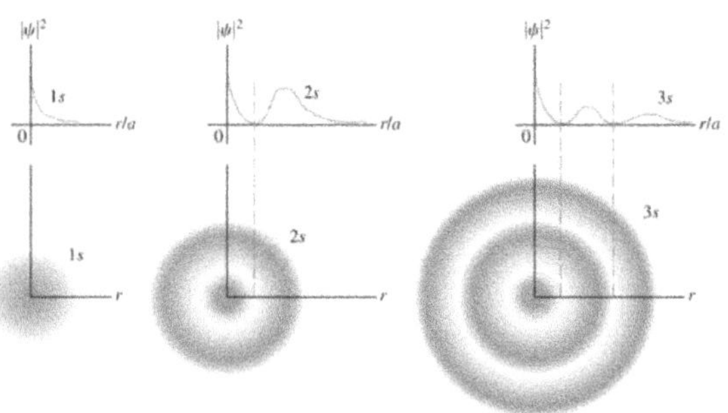

Figure 3.5: Probability distribution of electron in a Hydrogen atom

In quantum mechanics, the evolution of a quantum system is described by the Schrödinger equation, which dictates how the state of a system changes with time. The time operator in quantum mechanics is responsible for this evolution, and it acts on the state of a quantum system to produce a new state at a later time.

The probabilistic nature of quantum mechanics means that, even when we know the initial state of a quantum system and have a complete understanding of its governing equations, we can only predict the probability of outcomes at different times. This introduces an element of unpredictability and uncertainty into quantum systems that challenges our classical intuitions.

Quantum Time's Relevance

You might be wondering how these intricate aspects of time in the quantum world relate to our everyday experience. While it's true that the quantum behavior described here is most apparent at extremely small scales, the principles of quantum mechanics underpin the behavior of all matter and energy in the universe.

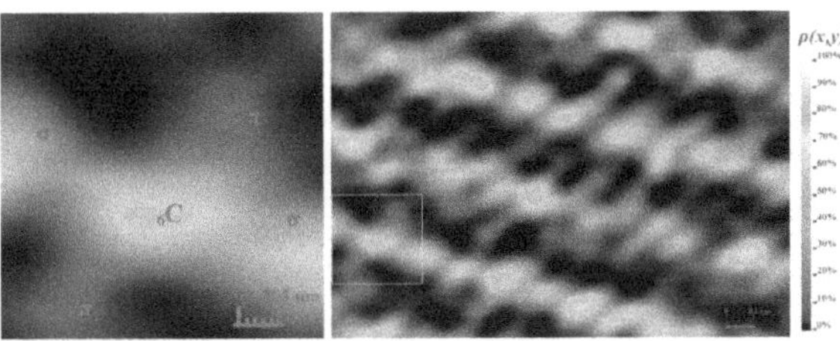

Figure 3.6: Photo of a carbon atom in graphite. Two inner electron clouds (pink color), two strong σ bonds (green) and two weak π bonds (blue) form the real shape of the carbon atom. Credit: Samsiq, CC BY-SA 4.0

For example, the behavior of atoms, molecules, and the particles within them is governed by quantum mechanics. These entities constitute the building blocks of our everyday reality, and their quantum nature is essential for understanding chemical reactions, materials science, and the behavior of matter and energy.

Additionally, quantum mechanics plays a pivotal role in the operation of modern technology, from semiconductors in our electronics to the lasers used in medical equipment. Without a deep understanding of quantum physics, the development of many of the technologies we rely on today would not have been possible.

Time's Quantum Mystery

The quantum world challenges our classical understanding of time, space, and reality. It introduces us to the concept of the quantum of time, where time is not a continuous and absolute dimension but rather a probabilistic and uncertain operator.

While the phenomena discussed here may seem far removed from our everyday experiences, they underpin the behavior of all matter and energy in the universe, including the technology we rely on and the processes that shape our world.

The enigmatic aspects of time in the quantum realm continue to inspire scientific inquiry, philosophical discussions, and imaginative explorations. As we move forward in our exploration of time, we'll journey even deeper into a spooky connection, mysteries of time travel, contemplate the cosmic implications of time, and ponder the uncharted territories of time's past and future.

Chapter 4

ENTANGLEMENT AND TIME: A SPOOKY CONNECTION

In the previous chapters, we journeyed through the history of timekeeping, explored the revolutionary concepts of relativity, and ventured into the enigmatic world of quantum mechanics. Now, we dive deeper into the heart of the quantum realm, where time and entanglement create a perplexing and captivating connection. The mysteries we'll uncover in this chapter challenge our classical intuitions and shed light on the intricate relationship between quantum entanglement and the nature of time.

Entanglement: A Quantum Connection

Entanglement is one of the most puzzling and fascinating phenomena in the quantum world. Imagine two entangled particles—let's call them Particle A and Particle B. If we measure a property of Particle A, such as its spin or polarization, the corresponding property of Particle B becomes instantaneously determined, regardless of the distance that separates them.

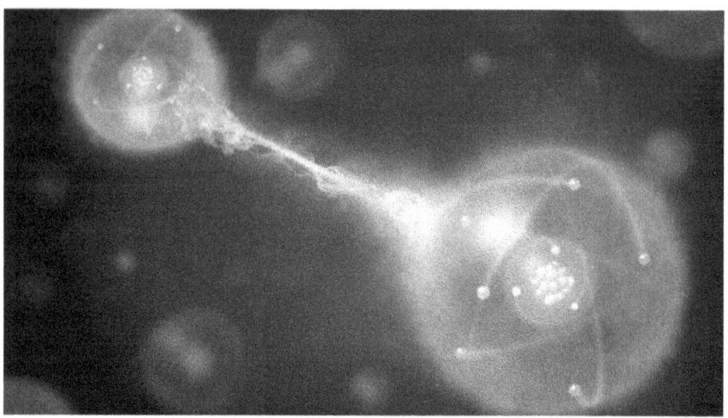

Figure 4.1: An artistic illustration of two entangled particles, with their properties correlated regardless of distance. Credit: Mark Garlick/Science Photo Library, Getty Images

Entanglement defies our classical intuitions about the separation of physical systems. In the classical world, the properties of objects are determined by their local interactions, and signals or influences cannot travel faster than the speed of light. However, entanglement demonstrates a form of instantaneous, nonlocal connection between particles.

This phenomenon was famously referred to by Albert Einstein as "spooky action at a distance" because it challenges our understanding of causality and the nature of reality. Although Einstein was skeptical of the implications of entanglement, numerous experiments have since confirmed its existence and its role in the quantum world.

The Einstein-Podolsky-Rosen (EPR) Paradox

The concept of entanglement was first introduced in a groundbreaking paper by Albert Einstein, Boris Podolsky, and Nathan Rosen in 1935, commonly known as the EPR paper. In this paper, the authors presented a thought experiment that highlighted the peculiar nature of entangled particles and the challenges it posed to the principles of quantum mechanics.

Albert Einstein   Boris Podolsky   Nathan Rosen

Figure 4.2: A photograph of (from left to right) Albert Einstein, Boris Podolsky, and Nathan Rosen

The EPR paradox involves two entangled particles, such as electrons, that are emitted from a common source. When one of these particles is measured, its properties, such as its spin or polarization, become known. The EPR paradox arises because, in the quantum world, measuring one of the entangled particles instantly determines the properties of the other particle, regardless of the distance that separates them.

This instant correlation, or "spooky action," appeared to violate the principle of locality, which states that physical processes in one location should not have immediate effects on distant locations. Einstein and his colleagues argued that quantum mechanics could not be a complete description of reality if it allowed such nonlocal connections.

Bell's Theorem and Experimental Verification

John Bell, a physicist, took the ideas of the EPR paradox further by formulating Bell's theorem. Bell's theorem provided a way to test the validity of Einstein's objections to quantum mechanics and the concept of entanglement.

Figure 4.3: A portrait of John Bell

Bell's theorem proposed that if the correlations between entangled particles exceeded certain bounds, it would be impossible to explain their behavior using any theory based on local hidden variables. In other words, if the predictions of quantum mechanics were upheld in experiments, it would demonstrate the reality of entanglement and challenge the principle of locality.

Subsequent experiments, known as Bell tests, confirmed the predictions of quantum mechanics and violated the limits set by local hidden variable theories. This provided strong evidence in favor of the nonlocal and entangled nature of quantum systems.

Time and Entanglement

The relationship between time and entanglement introduces intriguing questions about the nature of quantum reality. Entangled particles can exhibit instantaneous correlations, raising questions about the role of time in this process.

In classical physics, causality dictates that an effect cannot precede its cause, and the transfer of information should occur within a finite time. However, the instantaneous correlations observed in entangled particles defy these classical notions. If the properties of one entangled particle change instantly in response to a measurement of its partner, it seems as though the information is transmitted faster than the speed of light.

The resolution of this apparent paradox lies in the probabilistic nature of quantum measurements. While the properties of entangled particles may seem to change instantly, the specific outcomes of measurements are subject to probability. In other words, the correlations between entangled particles do not transmit information faster than the speed of light, and causality is preserved.

Quantum Time and Relativity

One of the most profound aspects of time in the quantum world is its relationship with the theory of relativity, as we explored in previous chapters. Special relativity, introduced by Albert Einstein, describes how the passage of time can vary depending on an observer's motion. When particles are entangled and move at different velocities relative to an observer, their experience of time can differ, which introduces unique considerations for the study of entanglement.

In special relativity, time dilation occurs when an object moves at a significant fraction of the speed of light. This effect means that time passes more slowly for a fast-moving object relative to a stationary observer. When particles are entangled and experience different velocities, their time dilation effects must be taken into account to understand their entangled behavior correctly.

Quantum Entanglement Experiments

Numerous experiments have been conducted to explore the phenomenon of quantum entanglement and its implications for the nature of time. These experiments confirm the reality of entanglement and the instantaneous correlations it produces.

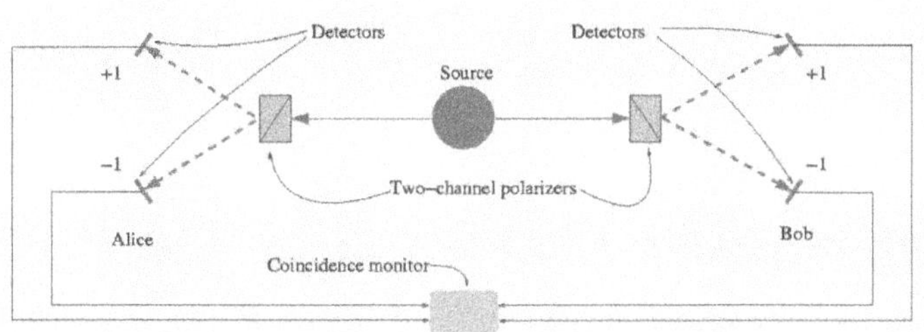

Figure 4.4: Schematic of Aspect experiment demonstrating quantum entanglement

One of the most famous experiments in this domain is the Aspect experiment, conducted by Alain Aspect in the 1980s. This experiment confirmed the violation of Bell's inequalities, providing strong evidence for the nonlocal nature of quantum systems.

Entanglement experiments continue to be a focus of research in quantum physics, offering insights into the profound connections between particles, space, time, and the nature of reality.

Quantum Entanglement and Quantum Computing

The phenomenon of entanglement has profound implications for the development of quantum computing. In classical computing, information is processed using bits, which can be either 0 or 1. In contrast, quantum computing uses quantum bits, or qubits, which can exist in multiple states simultaneously due to the principles of superposition and entanglement.

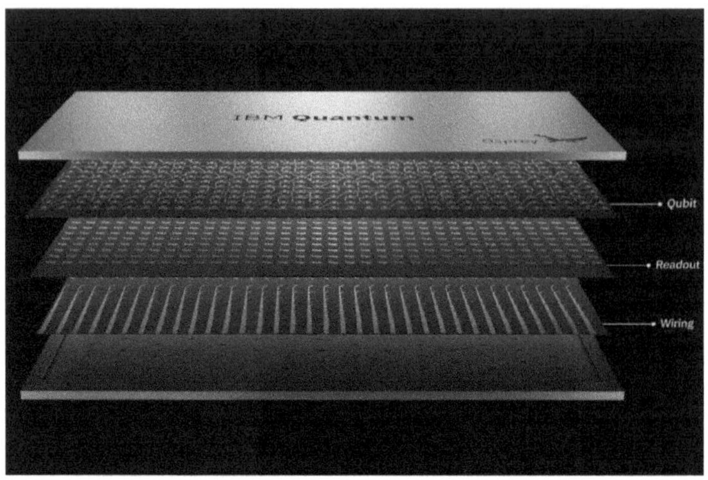

Figure 4.5: BM's 433-qubits Osprey quantum processor, with more than three times the qubits of the IBM Eagle processor unveiled in 2021. Credit: REUTERS

Entanglement plays a crucial role in the operation of quantum computers. By entangling qubits, quantum computers can perform complex calculations at speeds that surpass classical computers. This is because the correlations between entangled qubits allow them to share information and perform parallel computations, unlocking the potential for solving problems that are currently intractable for classical computers.

The development of quantum computing holds the promise of revolutionizing fields such as cryptography, optimization, and materials science, as well as providing solutions to problems previously thought to be unsolvable.

Entanglement and the Arrow of Time

The concept of entanglement introduces profound questions about the nature of time and the arrow of time. While the phenomenon itself involves instantaneous correlations, these correlations do not violate causality or the principles of the arrow of time.

In the quantum realm, time is treated differently from classical time. The probabilistic and uncertain nature of quantum measurements means that outcomes are determined through measurements and interactions. The instantaneous correlations observed in entanglement do not imply a violation of causality or a change in the arrow of time.

Entanglement serves as a reminder that the quantum world challenges our classical intuitions and highlights the richness and complexity of time in the quantum context. While it may appear mysterious and "spooky," the phenomenon of entanglement is a fundamental and experimentally confirmed aspect of quantum reality.

Implications for the Nature of Reality

The connection between entanglement and time raises deep philosophical questions about the nature of reality. The concept of entanglement implies that particles can exist in a state of superposition until measured, with their properties correlated instantaneously. This idea challenges our classical understanding of the deterministic, cause-and-effect nature of the universe.

One interpretation of entanglement suggests that it underscores the role of the observer in shaping reality. According to this view, the act of measurement collapses the superposition of quantum states into a definite outcome. This interpretation echoes the philosophical discussions initiated by the EPR paradox and the implications of quantum mechanics for the nature of time and causality.

Quantum Entanglement and Spooky Possibilities

Quantum entanglement is one of the most captivating and mysterious phenomena in the quantum world. Its instantaneous correlations, nonlocal connections, and implications for the nature of time continue to inspire scientific inquiry, philosophical discussions, and imaginative explorations.

While the concept of entanglement may seem far removed from our everyday experiences, it underpins the behavior of matter and energy at the quantum level, shaping the foundation of modern physics and technology. The phenomenon reminds us that reality is often more nuanced and intricate than our classical intuitions suggest.

As we delve deeper into the exploration of time in the chapters to come, we'll encounter even more intriguing and mind-expanding concepts. We'll contemplate the mysteries of time travel, ponder the cosmic implications of time, and journey through the uncharted territories of time's past and future. The quantum realm has revealed to us that time, like the universe itself, is far more complex, mysterious, and awe-inspiring than our classical intuitions could ever have imagined.

Chapter 5

TIME'S ARROW REVISITED

In our journey through the intricacies of time, we've explored the history of timekeeping, the groundbreaking theories of relativity, and the enigmatic world of quantum mechanics. Now, we turn our attention to a fundamental aspect of time—its directionality. Time, in our everyday experience, flows in a specific direction, moving from the past to the present and into the future. This characteristic of time, often referred to as the arrow of time, raises profound questions about the nature of the universe, the origin of our experiences, and the underlying physics that governs the passage of time. In this chapter, we revisit time's arrow, exploring its origins, the second law of thermodynamics, and its connection to entropy.

The Arrow of Time

Time, as we experience it, possesses a clear direction. We remember the past, live in the present, and anticipate the future. This inherent directionality is known as the arrow of time. It is one of the most fundamental aspects of our existence and shapes our perception of reality.

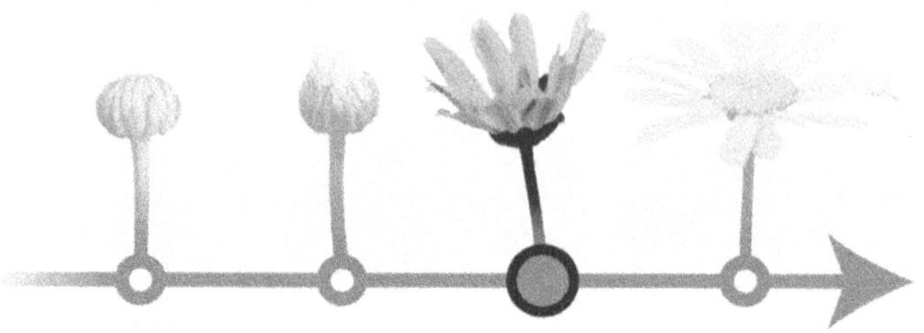

Figure 5.1: An illustration of the arrow of time, depicting the progression from past to present to future. Credit: Quanta Magazine

The concept of the arrow of time, also known as the "arrow of causality," is closely linked to the idea of causality itself. In our everyday lives, events occur in a particular order, and causes precede their effects. We don't witness teacups spontaneously shattering and gathering themselves back together, nor do we observe scrambled eggs unscrambling and returning to their unbroken state. The arrow of time gives us a sense of order and continuity in the universe.

The Symmetry of Physical Laws

One might expect the laws of physics to exhibit the same kind of temporal symmetry we encounter in our everyday lives. However, the fundamental laws of physics, such as Newton's laws of motion, Maxwell's equations for electromagnetism, and Einstein's equations of relativity, do not inherently distinguish between past and future.

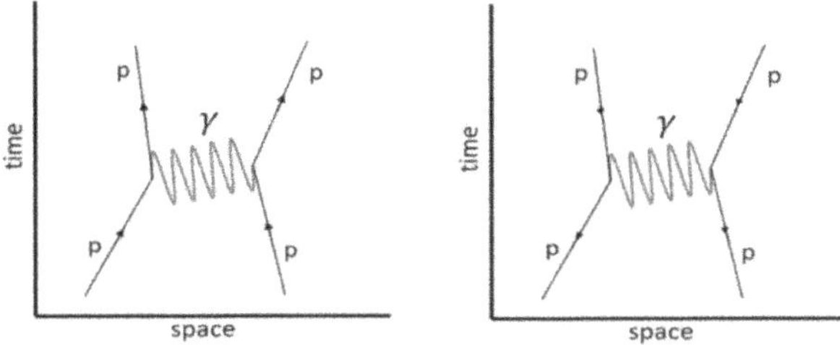

Figure 5.2: Feynman diagram illustrating the behavior of particles remains the same regardless of the direction of time

This property is known as time reversal symmetry. It implies that if we were to reverse the direction of time in a physical process while keeping all other conditions the same, the behavior of the system would remain unchanged. In other words, the fundamental equations that govern the behavior of particles and fields do not inherently enforce the arrow of time we perceive in our everyday lives.

To address this puzzle, we must explore the concepts of entropy and the second law of thermodynamics, which play a central role in understanding the arrow of time.

Entropy: The Measure of Disorder

In the 19th century, the physicist Rudolf Clausius introduced the concept of entropy as a measure of the degree of disorder in a system. Entropy is a fundamental concept in thermodynamics, which is the branch of physics that deals with the relationships between heat, work, energy, and the properties of matter.

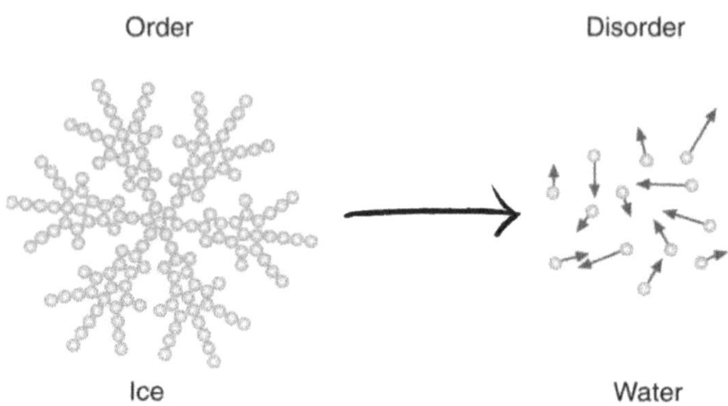

Figure 5.3: Image showing increase in entropy, changing the state from ice to water,

Entropy is often associated with the dispersal of energy and the increase in disorder in a closed system over time. In simple terms, it quantifies the amount of energy that is no longer available to do work in a system, having become too spread out or disorganized.

The Second Law of Thermodynamics

The second law of thermodynamics, formulated by Clausius, provides a fundamental principle for understanding the arrow of time. It states that in any isolated system, the total entropy always increases over time or remains constant, but it never decreases. In other words, the natural tendency of a closed system is to evolve toward a state of increasing disorder or entropy.

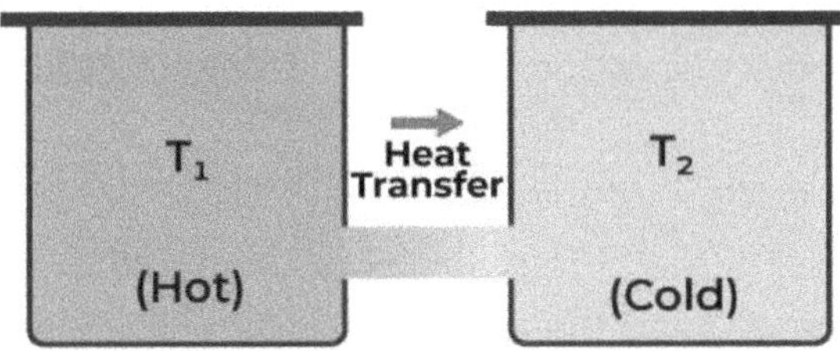

Figure 5.4: Second law of thermodynamics where entropy increases over time in an isolated system and achieves thermal equilibrium or the state of maximum entropy

This principle can be summarized in the famous phrase: "In any energy exchange, if no energy enters or leaves the system, the potential energy of the state will always be less than that of the initial state, increasing the entropy."

Entropy and the Arrow of Time

The second law of thermodynamics plays a critical role in the emergence of the arrow of time. It explains why we perceive time as flowing in a specific direction, from past to present and into the future. The arrow of time is intimately connected to the increase in entropy in our universe.

To illustrate this concept, consider a cup of hot coffee left on a table. Over time, the coffee cools down as heat dissipates into the surrounding environment. This cooling process is a manifestation of the second law of thermodynamics. Initially, the coffee and its heat energy were in a relatively ordered state, but as time progresses, the heat energy becomes more dispersed in the room, leading to an increase in entropy.

Figure 5.5: A depiction of a cup of hot coffee cooling down, illustrating the increase in entropy as heat disperses into the room. Credit: Victor De Schwanberg/Science Photo Library

The increase in entropy is a one-way process, leading to the perception of time's arrow. It explains why we do not observe coffee spontaneously reheating or heat energy spontaneously collecting itself back into the coffee cup. The arrow of time points in the direction of increasing entropy, from ordered to disordered states.

The Beginning of the Universe

One of the most intriguing questions related to the arrow of time is its origin and the conditions of the universe at its inception. The concept of entropy and the second law of thermodynamics have profound implications for our understanding of the early universe.

In the earliest moments of the universe, shortly after the Big Bang, the conditions were highly ordered, and entropy was at a minimum. As the universe expanded and evolved, entropy began to increase, leading to the emergence of galaxies, stars, and the complex structures we observe today.

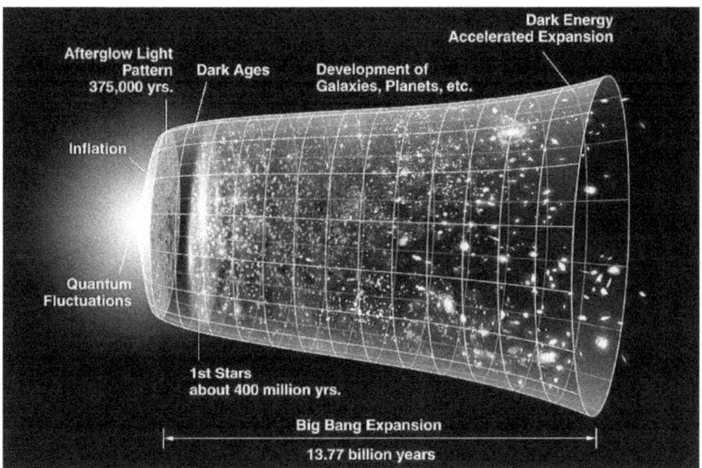

Figure 5.6: An artistic representation of the expansion of the universe from the Big Bang (highly ordered state) to present day (disordered state). Credit: NASA/WMAP Science Team

This scenario raises the question of why the universe began in a state of low entropy and has been progressing toward higher entropy. If the second law of thermodynamics holds, it implies that the universe's initial state was highly improbable, given the tendency of entropy to increase. This aspect of the arrow of time remains a subject of ongoing research and philosophical debate.

The Past and Future

The arrow of time has implications not only for the macroscopic world but also for our understanding of the past and future. In our daily lives, we can easily distinguish between past and future events based on the arrow of time. We remember the past, experience the present, and anticipate the future. However, in the realm of fundamental physics, the equations governing the behavior of particles do not inherently provide this distinction.

The fundamental laws of physics, whether classical or quantum, are time-reversible. This means that the equations work equally well whether time flows forward or backward. The concept of time reversal symmetry, as discussed earlier, illustrates that the behavior of particles and fields remains the same regardless of the direction of time.

So, how do we reconcile this time-reversible nature of fundamental laws with the irreversibility we observe in our everyday lives? The answer lies in the emergence of the arrow of time at macroscopic scales, driven by the second law of thermodynamics and the increase in entropy.

Time's Arrow and Consciousness

The arrow of time is not only a foundational concept in physics but also holds philosophical implications for the nature of consciousness and the human experience of time. Our perception of the past, present, and future is intricately linked to the arrow of time and the irreversibility of physical processes.

The philosopher Henri Bergson described two modes of time: "clock time," which is measurable and uniform, and "lived time," which encompasses our subjective experiences and the flow of our consciousness. The arrow of time is a central element of lived time, shaping our memories, experiences, and the sense of a continuous self.

Time's Mysteries

The arrow of time remains one of the most profound and mysterious aspects of our universe. While the fundamental laws of physics operate with time-reversibility, the everyday world introduces an inherent directionality, driven by the increase in entropy.

This paradox of time's arrow continues to inspire scientific inquiry, philosophical reflection, and creative exploration. The question of why the universe began with low entropy and is progressing toward higher entropy remains a topic of ongoing research and contemplation.

As we move forward in our journey through the mysteries of time, we'll explore the concept of time travel, contemplate the cosmic implications of time, and journey through the uncharted territories of time's past and future. The arrow of time, with its inherent directionality and the increase in entropy, offers us a unique perspective on the nature of our universe and our place within it. It invites us to ponder the enigmatic relationship between time, causality, and the unfolding of reality.

Chapter 6

TIME TRAVEL - FACT OR FICTION?

Time travel is a concept that has captivated human imagination for centuries. From H.G. Wells' "The Time Machine" to blockbuster movies like "Back to the Future," the idea of journeying through time has fascinated and inspired generations. But is time travel a real possibility, as suggested by science fiction, or is it forever confined to the realms of imagination? In this chapter, we'll explore the scientific and philosophical aspects of time travel and delve into the fascinating theories that surround it.

The Origins of Time Travel

The notion of time travel can be traced back to ancient mythology and folklore. Stories of time manipulation, temporal anomalies, and encounters with figures from the past or future have appeared in various cultures throughout history.

Figure 6.1: Time machine as imagined in H.G. Wells' classic novel. Credit: The Time Machine

One of the earliest recorded instances of time travel in literature can be found in the Hindu epic, the "Mahabharata," where the king Karna travels through time using celestial weapons. However, it was H.G. Wells who popularized the concept in his 1895 novella "The Time Machine," introducing the world to the idea of a machine that could transport individuals through time.

Wells' tale of a time-traveling adventurer who encounters the Eloi and the Morlocks in the distant future sparked the modern fascination with time travel. It also set the stage for a rich tapestry of time-travel narratives in science fiction.

Causality and the Grandfather Paradox

Causality is the principle that an event can cause another event, and it's a cornerstone of our understanding of time. In our everyday experience, we observe that cause and effect are connected in a linear, one-way fashion: the past influences the present, which in turn shapes the future.

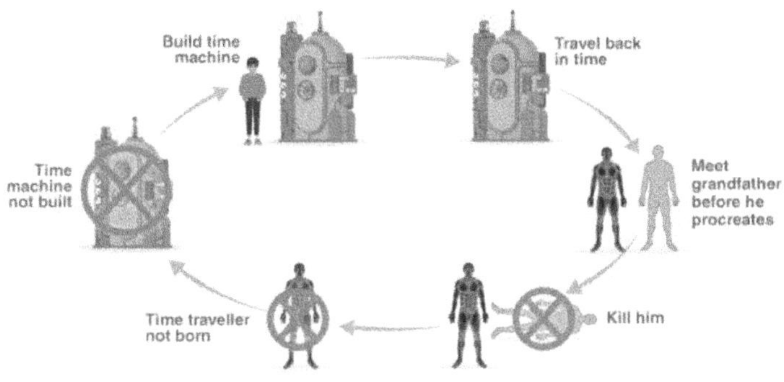

Figure 6.2: An illustration of the Grandfather Paradox. Credit: BYJUS

One of the most famous time travel paradoxes is the "Grandfather Paradox." It goes like this: If you were to travel back in time and prevent your grandparents from meeting, you would never have been born. If you were never born, you couldn't have traveled back in time to prevent your grandparents from meeting in the first place. This creates a logical contradiction that challenges the consistency of time travel.

The Grandfather Paradox and other causality dilemmas have been a topic of much discussion and debate among scientists and philosophers. They raise questions about whether time travel is fundamentally possible or if there might be some as-yet-unknown mechanism that prevents such paradoxes from occurring.

The Possibility of Time Travel

Now, let's explore the scientific and theoretical possibilities of time travel. While no one has built a functioning time machine, there are some intriguing concepts and theories within the realm of physics that suggest time travel could, in principle, be realized.

Special Relativity and Wormholes

Albert Einstein's theory of special relativity, which we discussed in Chapter 2, opens the door to the concept of time dilation. According to special relativity, time passes differently for objects in motion relative to each other. This leads to the intriguing possibility of "time dilation time travel."

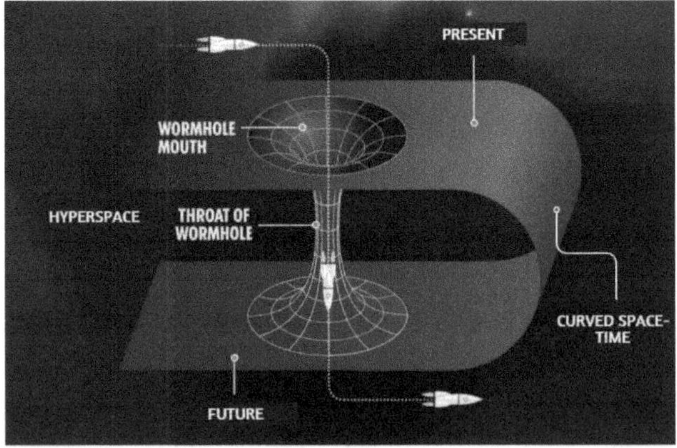

Figure 6.3: An artistic representation of a wormhole - a theoretical passage through spacetime

In the thought experiment we discussed in Chapter 2, a fast-moving spaceship could experience time passing more slowly than Earth-bound observers. This means that if an astronaut were to embark on a journey at near the speed of light and then return to Earth, they would find that less time has passed for them compared to people on Earth. Effectively, they would have traveled into the future.

While this is a form of time travel, it doesn't allow for going back to the past, as we saw in the Grandfather Paradox. This "one-way ticket to the future" scenario is supported by the principles of special relativity and has been experimentally confirmed.

The concept of wormholes, a speculative aspect of general relativity, offers another potential route to time travel. A wormhole is a theoretical tunnel-like structure that connects two separate points in spacetime. If one end of a traversable wormhole were to move at relativistic speeds and then return to its original location, time dilation effects could create a time difference between the two ends.

However, it's essential to understand that wormholes remain purely theoretical at this point. While they are a fascinating concept, there are significant challenges associated with their stability, creation, and maintenance.

General Relativity and Black Holes

Einstein's theory of general relativity, which we discussed in the same chapter, provides another avenue for exploring time travel, this time involving the extreme environments near black holes.

Figure 6.4: A snap from the movie Interstellar illustrating spaceship approaching a blackhole leading to time dilation. Credit: Interstellar

Near a black hole, the intense gravitational field causes significant spacetime curvature, leading to extreme time dilation effects. The closer one gets to a black hole, the slower time passes for them, as we discussed in the context of gravitational time dilation. This means that an object or observer approaching a black hole could experience time differently compared to those far away.

In theory, if an astronaut were to venture near a black hole and then return to a distant location, they might find that more time has passed on Earth than for themselves. This scenario is another form of time dilation time travel, but it, too, only allows for traveling to the future.

Cosmic Strings and Closed Timelike Curves

Cosmic strings are hypothetical objects predicted by some theories of the early universe. These strings are incredibly dense and thin, and they generate intense gravitational fields when they interact with matter. While cosmic strings remain speculative, they have led to discussions of the possibility of closed timelike curves (CTCs).

CTCs are paths through spacetime that loop back upon themselves, potentially allowing for time travel. These closed loops could enable an object or observer to return to an earlier point in their own timeline. However, CTCs also introduce paradoxes similar to the Grandfather Paradox, as changing events within a closed loop could create logical inconsistencies.

The existence of cosmic strings and CTCs is still a subject of debate and remains a highly speculative aspect of theoretical physics.

Quantum Time Travel

In the realm of quantum mechanics, the concept of time travel takes on a different character. While quantum mechanics introduces uncertainty and probabilistic behavior, it doesn't provide a straightforward mechanism for time travel.

The notion of parallel universes and the many-worlds interpretation has led to discussions of quantum time travel, where an individual could traverse between different branches of the multiverse, experiencing different outcomes. However, these concepts remain theoretical and have not been demonstrated.

The Reality of Time Travel

While the concepts we've discussed provide intriguing possibilities for time travel, it's important to emphasize that, as of now, time travel as depicted in science fiction remains in the realm of the imagination. The scientific and theoretical foundations that suggest time travel is possible are largely based on exotic and speculative aspects of physics.

Time travel would require a profound understanding of the laws of physics at scales and conditions we have yet to explore. It would also necessitate the development of technologies and mechanisms that are far beyond our current capabilities.

The paradoxes and challenges associated with time travel, such as the Grandfather Paradox and causality violations, have led some physicists to question whether time travel is fundamentally feasible or if there might be some underlying principle that prevents such scenarios from occurring.

The Philosophical Implications

Time travel raises profound philosophical questions about the nature of reality, causality, and free will. If time travel were possible, what would it mean for our sense of agency and responsibility? Could we change the past and shape our future at will?

The debate over the feasibility of time travel and the potential implications for our understanding of the universe and ourselves is ongoing. It challenges us to contemplate the boundaries of human knowledge and the mysterious nature of time itself.

As we journey through the universe of time, we'll encounter more mysteries, including the enigma of time's past and the cosmic implications of this profound dimension. Time travel may remain a tantalizing dream, but the pursuit of knowledge and the exploration of the unknown are at the heart of the human spirit.

Chapter 7

THE PHILOSOPHY OF TIME

In our exploration of time, we've journeyed through the annals of its measurement, navigated the relativistic realms, delved into the quantum mysteries that underlie its fabric, revisited arrow of time, and explored factuality of time travel. Now, as we approach the philosophical landscapes, we enter a realm where time transcends mere measurement and becomes a profound inquiry into the nature of existence. Welcome to the captivating realm of "The Philosophy of Time."

Time as a Philosophical Puzzle

The nature of time has long intrigued philosophers, sparking debates that transcend the boundaries of scientific inquiry. From ancient Greek thinkers to modern philosophers, questions about the nature of time, its reality, and its implications for our understanding of reality have permeated the corridors of intellectual discourse.

One of the earliest philosophers to contemplate the nature of time was Heraclitus, who famously declared, "You cannot step into the same river twice." This statement captures the essence of a philosophical puzzle: Is time a continuous flow, or is it composed of discrete moments? The river, with its ever-changing currents, serves as a metaphor for the dynamic and transient nature of reality.

A Brief History of Philosophical Perspectives on Time

Ancient Greek Philosophy

In ancient Greece, the concept of time found its place in the musings of philosophers such as Parmenides and Zeno. Parmenides, in his philosophical poem, asserted the unity and immutability of being, challenging the idea of change over time. Zeno, in a series of paradoxes, aimed to show the absurdity of motion and change, offering a set of puzzles that puzzled philosophers for centuries.

Augustine and the Eternal Present

Jumping forward to the 4th and 5th centuries, Saint Augustine of Hippo pondered the nature of time in his seminal work "Confessions." Augustine introduced the idea of the "eternal present," suggesting that past, present, and future are simultaneous in the mind of God. This perspective challenges the linear progression of time and presents a timeless, ever-present reality.

Figure 7.1: A portrait of Saint Augustine

Immanuel Kant's Transcendental Idealism

In the 18th century, Immanuel Kant brought a new perspective to the philosophical discourse on time with his theory of transcendental idealism. Kant argued that time is not an objective feature of the external world but a fundamental framework through which we perceive and organize our experiences. Time, according to Kant, is a subjective, a priori form of intuition that structures our perception of events.

Henri Bergson and Duration

As we enter the 20th century, Henri Bergson contributed significantly to the philosophy of time with his concept of "duration." Bergson distinguished between two types of time: measured, quantitative time (clock time) and lived, qualitative time (duration). Duration, according to Bergson, is a continuous flow of subjective experience, capturing the richness and depth of our lived moments.

Figure 7.2: A portrait of Henri Bergson

Philosophers throughout history have grappled with the fundamental nature of time. Is time an objective reality that exists independently of our perception, or is it a subjective construct shaped by our consciousness? The exploration of these questions opens the door to different philosophical viewpoints on the essence of time.

Eternalism vs. Presentism

Two major philosophical positions regarding the nature of time are eternalism and presentism. Eternalism asserts that past, present, and future are all equally real and exist concurrently. This perspective views time as a block universe, where every moment in time has a fixed existence. Presentism, on the other hand, contends that only the present moment is real, and the past and future are mere abstractions.

The Experience of Time

While philosophers debate the metaphysical nature of time, the subjective experience of time remains a central theme. Our perception of time is deeply intertwined with our consciousness, and how we subjectively experience the passage of time has profound implications for our understanding of temporal reality.

The Arrow of Subjective Time

The arrow of subjective time refers to our internal sense of the flow of time, from past to present to future. This arrow is intimately connected to our experiences, memories, and anticipations, shaping our unique journey through the temporal landscape. The subjectivity of time challenges us to consider how our perceptions influence the broader understanding of temporal reality.

The Block Universe: A Static Vision of Time

Eternalism paints a picture of time as a static and unchanging block, where past, present, and future coexist. This perspective challenges our intuitive sense of a dynamic and evolving reality, urging us to reconsider the nature of free will, causality, and the very essence of change.

Special Relativity and the Block Universe

Einstein's theory of special relativity aligns with the eternalist view of a block universe. In the spacetime framework introduced by special relativity, events are positioned in a four-dimensional continuum, challenging the traditional linear concept of time.

Implications for Free Will

The block universe model raises philosophical questions about free will and determinism. If the future already exists in its entirety, do we have the freedom to shape our destinies, or are our choices predetermined by the unchanging block of time? The interplay between free will and determinism in the context of the block universe remains a topic of philosophical discourse.

The Dynamic Present: Presentism Reconsidered

While eternalism presents a static vision of time, presentism argues for the dynamic nature of the present moment. The present, in this view, is a unique and ever-changing instant that holds a special ontological status.

The Now and the Flow of Time

Presentism emphasizes the significance of the "now," suggesting that the present moment is the only real and existing component of time. The flow of time, according to presentism, is an ongoing and dynamic process, with each moment unfolding into the next.

Temporal Becoming and Change

The philosophy of temporal becoming, associated with presentism, underscores the idea that reality is in a constant state of change and becoming. This perspective resonates with our intuitive sense of time as a flowing river, with each moment giving way to the next.

Time and Existential Reflections

The philosophy of time extends beyond abstract inquiries into metaphysics and causality; it also invites us to reflect on the nature of existence and our place in the temporal tapestry.

Existential Temporality

Existentialist philosophers, such as Jean-Paul Sartre and Albert Camus, contemplated the temporal aspects of human existence. Sartre, in his work "Being and Nothingness," emphasized the concept of "bad faith," where individuals deny their freedom and responsibility by succumbing to societal expectations and routines.

Figure 7.3: Portraits of Jean-Paul Sartre (left) and Albert Camus (right)

Camus, in "The Myth of Sisyphus," explored the existential condition of living in a seemingly indifferent universe. The myth of Sisyphus, condemned to roll a boulder uphill for eternity, serves as a metaphor for the human struggle to find meaning and purpose in the face of the repetitive and sometimes absurd nature of life.

Time and the Aesthetics of Existence

The philosophy of time also intersects with the aesthetics of existence—the ways in which we appreciate and engage with the beauty, impermanence, and transience of life.

Aesthetic Temporality

The German philosopher Friedrich Nietzsche, known for his profound insights into the human condition, explored the aesthetic aspects of existence. Nietzsche introduced the concept of the "eternal recurrence," suggesting that we should live our lives as if we would have to relive them over and over again.

Figure 7.4: A portrait of Friedrich Nietzsche

Nietzsche's idea challenges us to approach life with a heightened sense of appreciation, recognizing the ephemeral nature of moments and the potential for profound beauty in our experiences. It invites us to embrace the fleeting nature of time and infuse our lives with meaning and creativity.

Time and the Unanswerable Questions

As we navigate the philosophical terrain of time, we encounter questions that transcend easy answers and probe the boundaries of human understanding.

The Mystery of Temporal Origins

The question of temporal origins remains a philosophical and scientific mystery. What initiated the unfolding of time itself? Was there a temporal beginning, or does time extend infinitely into the past? These questions transcend the limits of our current knowledge and delve into the mysteries of existence.

Time and Ultimate Reality

Philosophers have pondered whether time is a fundamental aspect of ultimate reality or merely a human construct. Does time exist independently of our perception and measurement, or is it a product of our cognitive processes and cultural frameworks?

This inquiry into the ultimate nature of time invites us to consider the relationship between our conceptualizations of time and the underlying fabric of reality. It challenges us to confront the limitations of human understanding in the face of cosmic mysteries.

Time and Human Experience

In our exploration of the philosophy of time, we've encountered a tapestry woven with threads of metaphysical speculation, existential reflections, and aesthetic contemplation. Time, in its enigmatic and multifaceted nature, permeates the very essence of human experience.

As we continue our journey through the mysteries of time, we'll explore the cosmic implications of time, and venture into the uncharted territories of time's past and future.

Chapter 8

THE COSMOS AND TIME

As we journey deeper into the realms of time, our gaze extends beyond the confines of Earth, reaching toward the cosmos—a vast expanse where the dance of celestial bodies unfolds in the grand tapestry of space and time. In this chapter, we embark on a cosmic journey through the universe, delving into the birth of the cosmos, the vastness of cosmic time, and the unfolding narrative of the cosmos.

Time's Dance in the Cosmos

In the cosmic ballet, time takes center stage as a fundamental player, orchestrating the movements of stars, galaxies, and the very fabric of space itself. To understand the cosmic dimensions of time, let's embark on a journey through the milestones that mark the cosmic timeline.

The Birth of the Cosmos: The Big Bang

Our cosmic story begins approximately 13.8 billion years ago with an event of unparalleled significance—the Big Bang. In a fraction of a second, the universe expanded from an infinitesimally small, hot, and dense state to a vast and evolving cosmos. This cosmic inception marked the beginning of time as we know it.

The concept of the Big Bang might sound like an explosive event, but it wasn't an explosion in a pre-existing space. Instead, it represents the expansion of space itself. As the universe expanded, it carried with it the seeds of matter, energy, and the cosmic symphony of time.

The Cosmic Symphony: Galaxies, Stars, and Planets

As the universe matured, matter coalesced into vast structures known as galaxies. These cosmic cities, comprising billions to trillions of stars, became the theaters where the narrative of time unfolded on a grand scale.

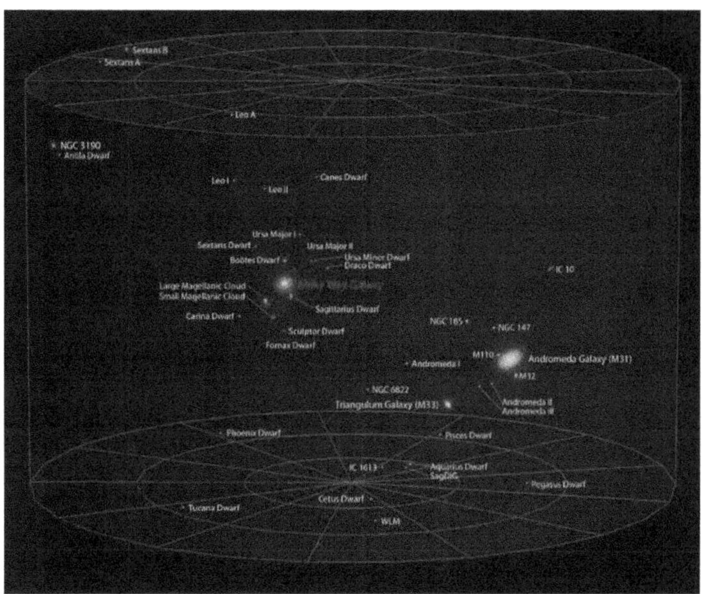

Figure 8.1: An artistic representation of the Local Galactic Group. Credit: Andrew Z. Colvin, CC BY-SA 3.0

Within galaxies, stars ignited, radiating light and energy into the cosmic expanse. These stellar actors experienced their own lifecycle, from birth in stellar nurseries to dramatic finales, such as supernova explosions. The eons passed as stars danced in cosmic time, creating the elements that would later become the building blocks of planets and life.

In the outskirts of galaxies, planetary systems formed, each with its own celestial bodies tracing orbits around a central star. On one such planet, Earth, the stage was set for the emergence of life and the development of intelligent beings capable of contemplating the mysteries of time.

The Cosmic Calendar: A Year in the Life of the Universe

To grasp the enormity of cosmic time, let's compress the entire history of the universe into a single calendar year, known as the Cosmic Calendar. In this cosmic chronicle, the Big Bang occurs at midnight on January 1st, and each month represents approximately 1.17 billion years.

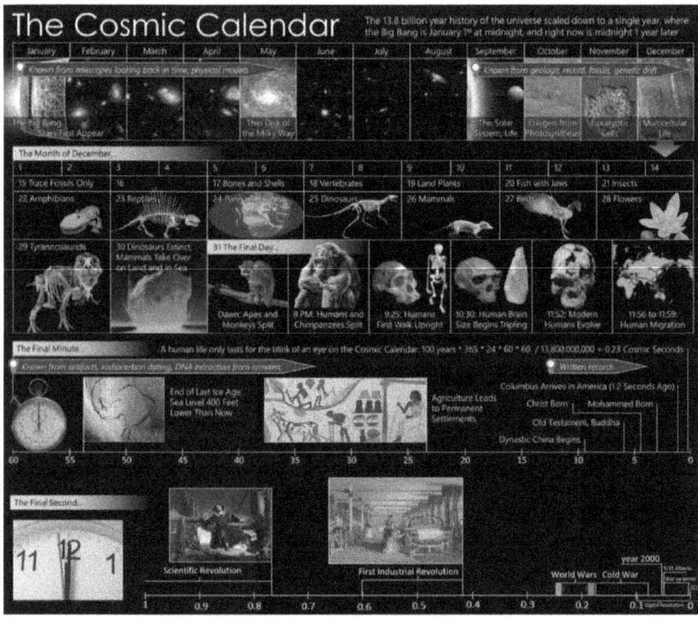

Figure 8.2: A graphical view of the Cosmic Calendar, featuring the months of the year, days of December, the final minute, and the final second. Credit: Efbrazil, CC BY-SA 3.0

- January 1st: The Big Bang marks the beginning of the universe.

- January 26th: The first galaxies form.

- September 2nd: Our solar system, including Earth, takes shape.

- September 21st: Life emerges on Earth.

- December 25th: Dinosaurs roam the Earth.

- December 31st: Homo sapiens appear, marking the last minutes of the cosmic year.

This compressed timeline underscores the vast stretches of time preceding the emergence of humanity. It humbles us by revealing that the events of human history occupy mere moments in the cosmic calendar.

Cosmic Clocks and the Cosmic Microwave Background

In our exploration of cosmic time, we encounter a celestial timekeeper that offers insights into the early moments of the universe—the Cosmic Microwave Background (CMB). The CMB is a faint glow of radiation permeating the cosmos, originating from a time when the universe transitioned from an opaque, hot state to a transparent, cooler era.

Echoes of the Big Bang: The Cosmic Microwave Background

Approximately 380,000 years after the Big Bang, the universe had cooled sufficiently for electrons and protons to combine, forming neutral hydrogen atoms. This cosmic transition allowed photons to travel freely through space, creating the CMB—a snapshot of the universe's early moments.

Figure 8.3: Cosmic Microwave Background radiation from Wilkinson Microwave Anisotropy Probe (WMAP). Credit: NASA/WMAP Science Team

The CMB serves as a cosmic time capsule, preserving information about the density fluctuations and conditions of the early universe. Studying the patterns in the CMB provides astronomers with valuable clues about the cosmic ingredients that shaped the formation of galaxies and cosmic structures.

Cosmic Soundwaves: Acoustic Oscillations in the Early Universe

To understand the origin of the CMB's patterns, imagine the early universe as a cosmic symphony. In its infancy, soundwaves propagated through the primordial plasma, creating regions of compression and rarefaction.

As the universe expanded and cooled, these acoustic oscillations left their imprint on the distribution of matter.

When we observe the CMB, we are listening to the cosmic echoes of these primordial soundwaves, providing a unique window into the early chapters of cosmic history.

The Expanding Universe

Embarking on its cosmic odyssey, the universe unfolds as a dynamic, ever-expanding tapestry, challenging our perception of cosmic time. At the heart of this revelation is the phenomenon of cosmic redshift - where light from distant galaxies appears slightly shifted toward the longer-wavelength, or "red," end of the spectrum, a cosmic signature that beckons us to explore the underlying dynamics of the cosmos.

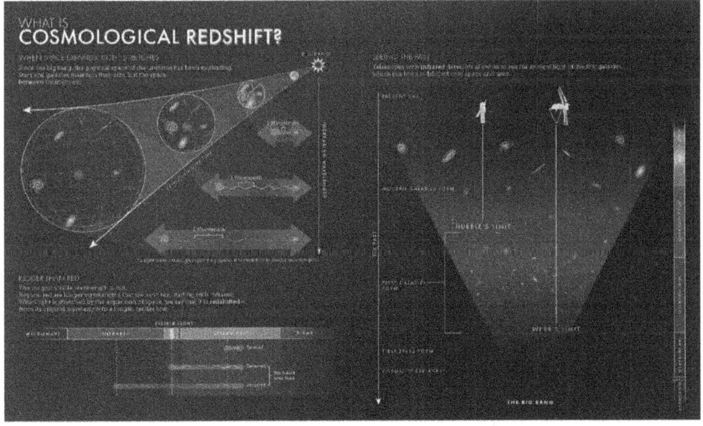

Figure 8.4: cosmological redshift, or the stretching of light waves due to the expansion of space between galaxies. Credit: NASA, ESA, Leah Hustak (STScI)

Hubble's Law: Velocity and Distance

In the annals of astronomical breakthroughs, Edwin Hubble's pioneering observation in 1929 shattered the notion of a static universe. He uncovered a celestial dance where galaxies, like cosmic dancers, gracefully move away from each other. This cosmic ballet is encapsulated in Hubble's law, an elegant equation that intertwines the velocity of galactic recession with their distance from our vantage point. The implications are profound, unfurling a narrative of cosmic evolution written in the celestial distances.

Figure 8.5: A photograph of Edwin Hubble

The Scale of the Universe

To fathom the vastness of cosmic time, one must grapple with the incomprehensible scale of the universe. From the minutiae of individual galaxies to the grandeur of galactic clusters and the cosmic voids that separate them, the celestial canvas spans unimaginable expanses. Studying this cosmic scale invites contemplation of the cosmic majesty, humbling us before the celestial magnificence that unfolds across colossal distances.

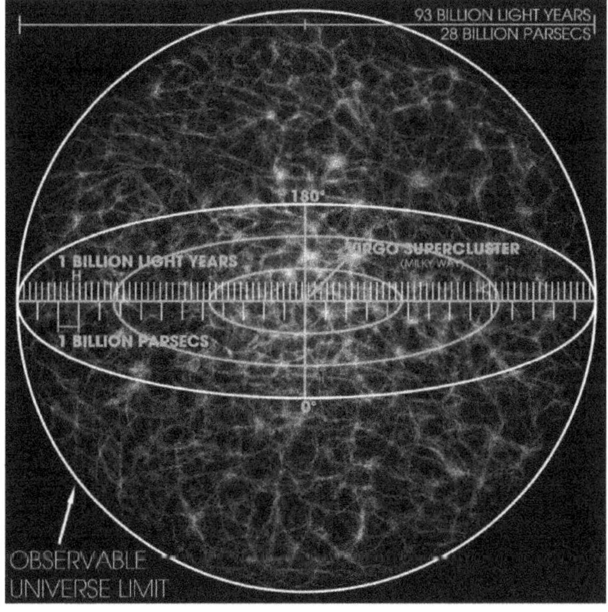

Figure 8.6: A visual representation of the scale of the universe, showcasing galaxies, clusters, and voids. Credit: Andrew Z. Colvin, CC BY-SA 3.0

The End of Time: Heat Death or Something Else

As our journey through cosmic time progresses, the narrative inevitably turns towards the contemplation of the universe's ultimate fate. Will the cosmic symphony continue its melodic expansion into eternity, or does the universe face a cold and desolate conclusion known as the heat death? The cosmic script tantalizes us with uncertainties and alternative possibilities that challenge our conventional understanding of the end of time.

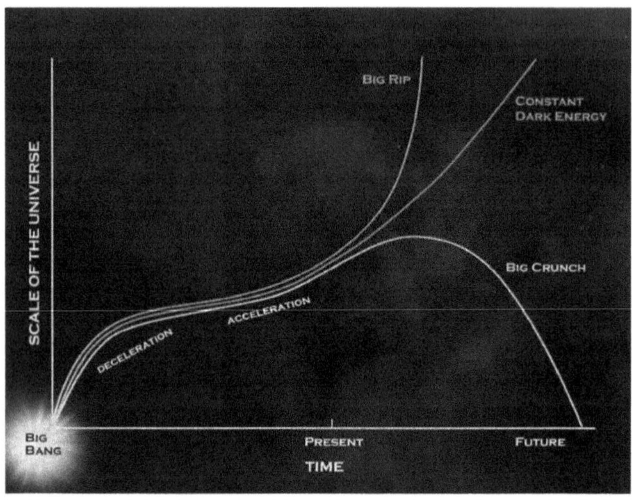

Figure 8.7: The far distant fates of the Universe offer a number of possibilities, but if dark energy is truly a constant, as the data indicates, it will continue to follow the red curve, leading to the long-term scenario described here: of the eventual heat death of the Universe. However, dark energy doesn't have to be a cosmological constant. Credit: NASA/CXC/M. Weiss

Heat Death: Entropy's Cosmic Destiny

The ominous concept of the heat death casts a shadow over the cosmic horizon, grounded in the inexorable rise of entropy. As the universe expands, energy gradients dissipate, and systems inexorably approach thermodynamic equilibrium. A point where even stars, like our sun, and black holes start to fade away, inching towards a state of maximum entropy and evaporating into the cosmic mist. The arrow of time points inexorably towards an eventual cosmic stillness, a celestial hush as the universe succumbs to the clutches of entropy.

Alternatives and Cosmic Revitalization

Yet, amid the cosmic murmur of the heat death, alternative hypotheses emerge as cosmic whispers challenging the presumed inevitability of cosmic demise. Enter the enigmatic realms of the Big Bounce, where the universe experiences cyclical oscillations of expansion and contraction, or the notion of a cyclic universe perpetually renewing itself. These alternative perspectives extend beyond the linear narrative of a one-way journey towards heat death, painting a cosmos resilient in its ability to redefine and revitalize its cosmic destiny.

The Ever-Unfolding Cosmos

The exploration of cosmic time, from the birth of cosmos to the dance of galaxies to the contemplation of cosmic endings, invites us to journey through the cosmic epochs. As we gaze into the celestial abyss, the mysteries of big bang, redshift, cosmic expansion, and the cosmic destiny unfold like chapters in an epic novel, enticing us to embark on an odyssey of cosmic exploration and understanding. The cosmos, in its vastness, holds the promise of endless discovery, challenging our intellect and inspiring our imagination in the quest to comprehend the timeless expanse of the universe.

Chapter 9

TIME IN TECHNOLOGY

In the ever-evolving landscape of human innovation, the intersection of time and technology shapes the present and charts the course for a future where the boundaries between science fiction and reality blur. In this chapter, we embark on a journey through the digital realms, where the ticking of clocks echoes in the algorithms of artificial intelligence, the precision of atomic timekeeping, and the transformative potential of emerging technologies.

From Atomic Clocks to Precision Timekeeping

Our exploration of time in technology begins with the unsung heroes of precision timekeeping—the atomic clocks. These remarkable instruments, operating on principles rooted in quantum mechanics, redefine our ability to measure time with unprecedented accuracy.

The Atomic Ballet: Quantum Precision

The transition of electrons within atoms serves as the metronome for atomic clocks. In the realm of cesium atomic clocks, the oscillations of cesium atoms determine the standard unit of time—the second. More recently, optical lattice clocks, utilizing transitions in strontium or ytterbium atoms, have elevated precision to astonishing levels.

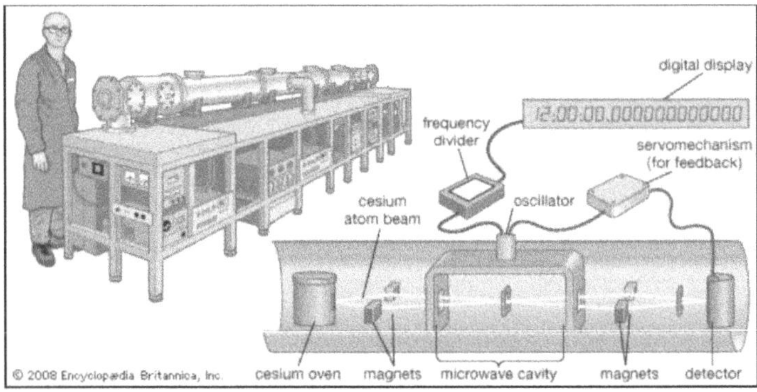

Figure 9.1: An illustration depicting the functioning of caesium atomic clock. Credit: Encyclopaedia Britannica

To grasp the precision of these timekeepers, imagine an atomic clock that would lose or gain only one second over the age of the universe—approximately 13.8 billion years. Such extraordinary accuracy not only benefits scientific research but also plays a crucial role in modern technologies, including global positioning systems (GPS), telecommunications, and financial transactions.

Atomic Timekeeping in Everyday Life

Atomic timekeeping has seamlessly integrated into our daily routines, ensuring the synchronization of devices and systems that rely on precise timing. From the coordination of satellite networks that power GPS navigation to the accurate timestamping of financial transactions in global markets, atomic clocks underpin the technological infrastructure that shapes the interconnected world.

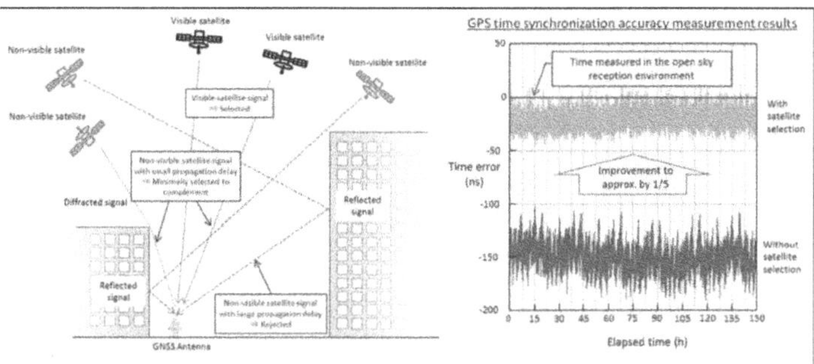

Figure 9.2: Satellite selection algorithm and GNSS receiver prototype performance test results. Credit: FURUNO ELECTRIC

As we navigate the digital landscape, the invisible heartbeat of atomic timekeeping resonates in the background, facilitating the seamless functioning of the technologies we often take for granted.

Artificial Intelligence and Machine Learning

In the realm of technology, the marriage of time and algorithms gives rise to the transformative power of artificial intelligence (AI) and machine learning. These digital entities, driven by complex algorithms, redefine our relationship with time in the domains of data analysis, decision-making, and predictive modeling.

Learning from the Past: Predictive Analytics

The ability of AI systems to analyze vast datasets and recognize patterns opens new frontiers in predictive analytics. By learning from historical data, these algorithms can forecast future trends, anticipate user preferences, and optimize a myriad of processes.

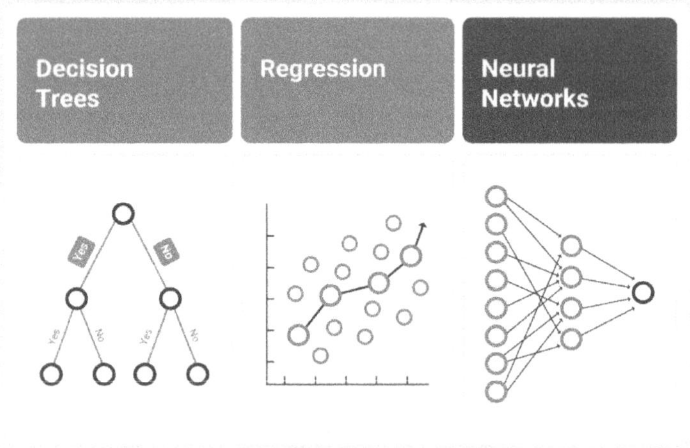

Figure 9.3: A few predictive algorithms in AI

Consider the recommendations offered by streaming platforms, suggesting movies or music based on past preferences. Behind these suggestions, AI algorithms analyze user behavior, discern patterns, and predict potential interests, creating a personalized and time-efficient experience.

Real-Time Decision-Making: AI in Action

In sectors such as finance, healthcare, and cybersecurity, AI systems operate in real-time, making split-second decisions that impact outcomes. These decision-making processes leverage the ability of algorithms to rapidly process information, identify anomalies, and adapt to dynamic situations.

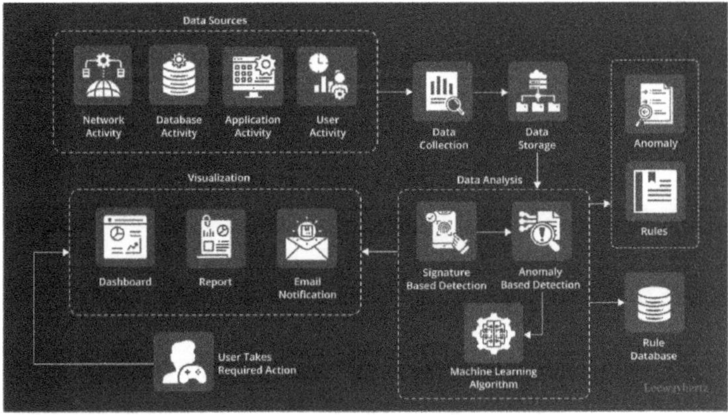

Figure 9.4: A schematic of real-time decision-making capabilities of AI in cybersecurity. Credit: LeewayHertz

For example, in the realm of cybersecurity, AI algorithms analyze network traffic patterns in real-time, swiftly detecting and responding to potential threats before they escalate. The integration of AI-driven decision-making enhances the resilience and adaptability of technological systems.

Quantum Computing on the Horizon

As we peer into the future, the realm of quantum computing emerges as a frontier that challenges our conventional understanding of time and computation. Quantum computers harness the principles of quantum mechanics to process information in ways that defy the limitations of classical computing.

Quantum Bits: Entanglement and Superposition

At the heart of quantum computing are quantum bits, or qubits. Unlike classical bits, which exist in a state of

0 or 1, qubits can exist in a superposition of both states simultaneously. This inherent duality, coupled with the phenomenon of entanglement, allows quantum computers to perform parallel computations on an exponential scale.

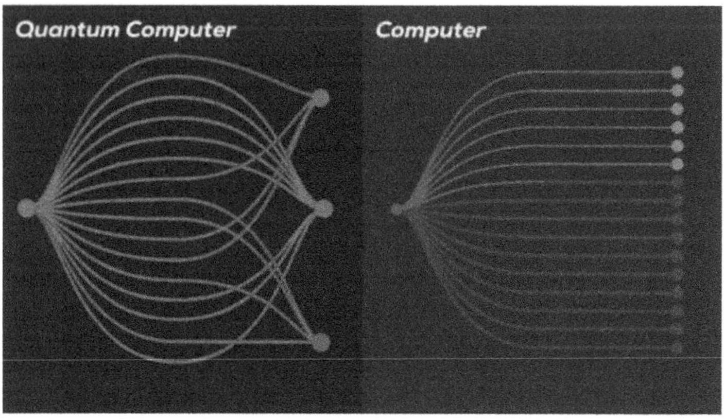

Figure 9.5: An illustration depicting the principles of entanglement and superposition in quantum computing leading to measuring result exponentially more efficient than a normal computer. Credit: Kurzgesagt

Entanglement enables qubits to share information instantaneously, regardless of the distance between them. This interconnectedness opens the door to the creation of quantum computers capable of solving complex problems exponentially faster than classical computers.

Quantum Supremacy: A Glimpse of the Future

In 2019, Google's quantum computer, Sycamore, achieved a milestone known as quantum supremacy. This term denotes the point at which a quantum computer outperforms the most advanced classical supercomputers in a specific task.

Figure 9.6: Google's quantum computer, Sycamore, which achieved quantum supremacy in 2019. Credit: Google's quantum AI lab

Google's achievement marked a significant leap forward in the quest for practical quantum computing. While quantum supremacy in itself does not guarantee the immediate resolution of complex problems, it signals the potential for quantum computers to revolutionize fields such as cryptography, optimization, and materials science in the years to come.

The Internet of Things (IoT)

In the interconnected landscape of the present and the future, the Internet of Things (IoT) emerges as a paradigm that seamlessly integrates time into the fabric of our daily lives. The IoT represents a network of interconnected devices, sensors, and systems that communicate and collaborate, creating a web of real-time data exchange.

Smart Homes and Wearables: Living in Sync with Time

The integration of IoT devices into our homes transforms living spaces into smart environments that respond to our needs in real-time. From thermostats that adjust based on occupancy patterns to wearables that monitor health metrics, these devices create a dynamic ecosystem attuned to the rhythms of daily life.

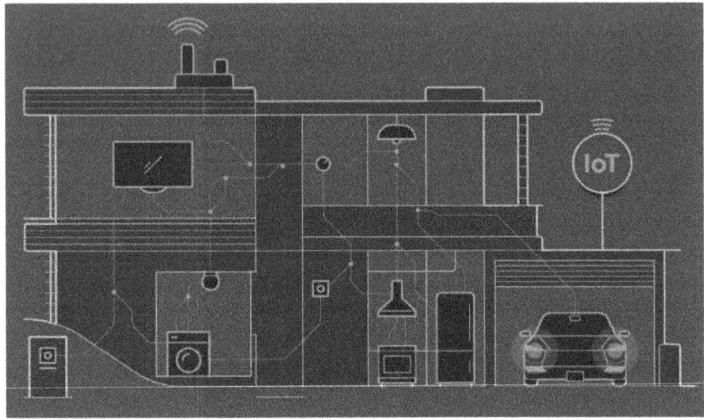

Figure 9.7: An illustration depicting the integration of IoT devices in a smart home. Credit: Toptal

Consider the scenario of a smart home where lights automatically adjust brightness based on the time of day, or a wearable device that tracks the wearer's sleep patterns and adjusts the alarm accordingly. These real-time adaptations enhance efficiency, comfort, and overall well-being.

Industrial IoT: Optimizing Processes and Efficiency

In the industrial landscape, the Industrial Internet of Things (IIoT) revolutionizes manufacturing and production processes. IoT-enabled sensors and devices collect real-time data, providing insights that drive efficiency, predictive maintenance, and resource optimization.

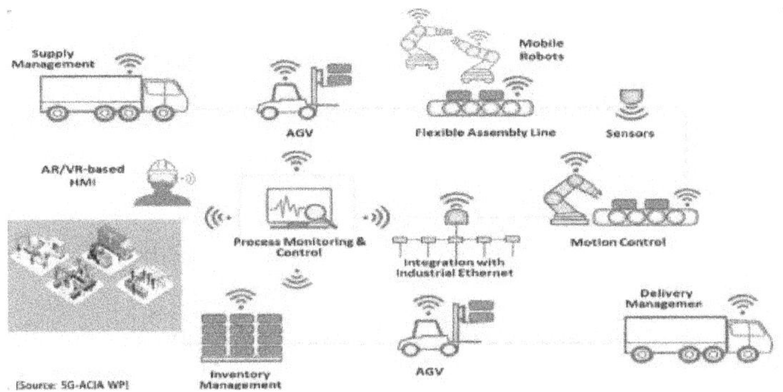

Figure 9.8: IoT in 6G based smart factory of the industry 4.0 system.
Credit: Neural Computing and Applications, SpringerLink

For example, in manufacturing plants, sensors monitor equipment performance, detect anomalies, and trigger maintenance actions before issues escalate. This proactive approach minimizes downtime, reduces costs, and enhances the overall efficiency of industrial operations.

Quantum Cryptography and Cybersecurity

As technology advances, the question of secure communication becomes increasingly crucial. Quantum cryptography emerges as a frontier that harnesses the principles of quantum mechanics to address the challenges of secure communication in the digital age.

Quantum Key Distribution: Unbreakable Encryption

Quantum Key Distribution (QKD) leverages the principles of quantum mechanics to secure communication channels with unbreakable encryption. In traditional cryptography, the security of encrypted messages relies on the complexity of mathematical algorithms, which could be vulnerable to advancements in computing power.

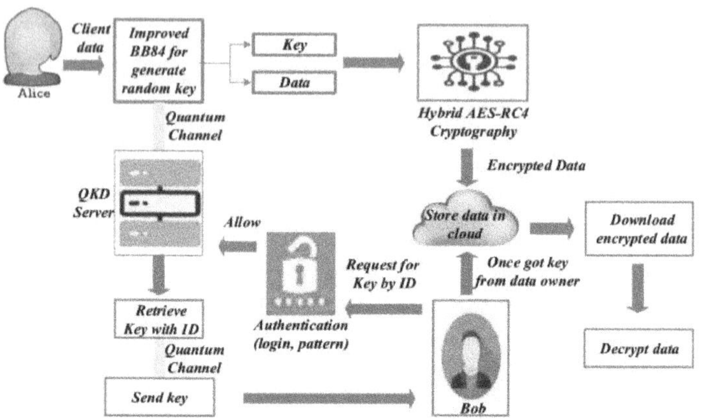

Figure 9.9: An architecture of Quantum Key Distribution for secure communication. Credit: Wireless Personal Communications, SpringerLink

QKD, on the other hand, utilizes the quantum properties of particles, such as photons, to create cryptographic keys that are intrinsically secure. Attempts to intercept these quantum keys would disrupt the quantum state, alerting the communicating parties to potential eavesdropping.

Time and Emerging Technologies

As we navigate the present landscape of technology, glimpses of the future beckon us toward transformative possibilities. Emerging technologies on the horizon promise to reshape the way we perceive and interact with time, pushing the boundaries of innovation to new frontiers.

Quantum Internet: Entangled Connectivity

The concept of a quantum internet envisions a global network where quantum entanglement facilitates secure and instantaneous communication. Quantum internet protocols, currently in the experimental stage, aim to create a communication infrastructure that leverages the quantum properties of particles for enhanced security and connectivity.

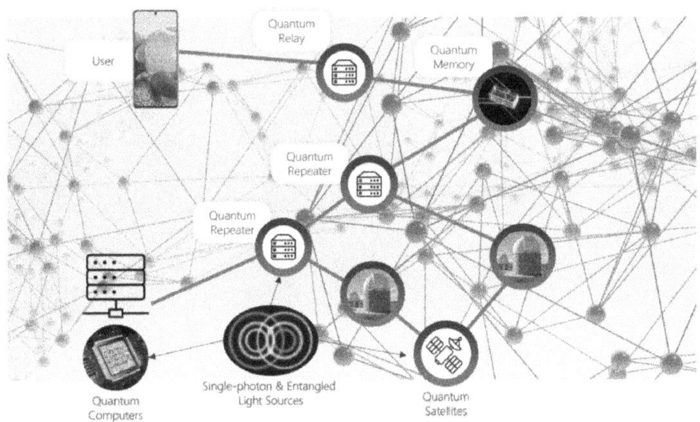

Figure 9.10: Topology of the Quantum Internet. Credit: ID Quantique

In a quantum internet, entangled particles could be used to establish connections that transcend traditional limitations. The prospect of a quantum internet hints at a future where secure communication, quantum computing, and quantum cryptography converge to create a new paradigm in the digital realm.

Brain-Computer Interfaces: Merging Minds with Machines

The frontier of brain-computer interfaces (BCIs) holds the promise of direct communication between the human brain and digital systems. BCIs aim to decode neural signals, enabling individuals to interact with devices, control prosthetics, and even communicate with each other through thoughts.

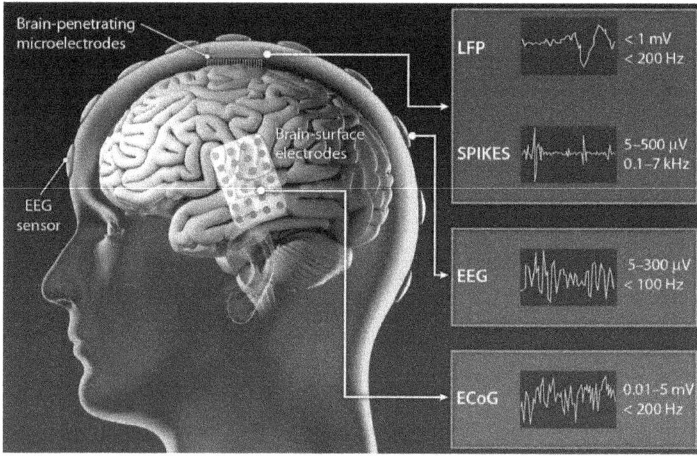

Figure 9.11: An illustration depicting the concept of a brain-computer interface. Credit: Nicolette Driscoll

In the realm of time, BCIs could revolutionize the way we perceive and manipulate temporal information. Imagine the ability to navigate through digital interfaces, manipulate time-sensitive tasks, or communicate with others through the seamless integration of neural signals and technological systems.

Navigating the Temporal Frontiers

As we conclude our journey through "Time in Technology," we find ourselves at the nexus of the present and the unfolding future. The digital landscapes of atomic timekeeping, artificial intelligence, quantum computing, and the Internet of Things beckon us to explore the temporal frontiers that shape our technological reality.

In this digital era, time becomes more than a measure on a clock; it transforms into a dynamic force that influences the way we connect, innovate, and envision the future. As we stand on the precipice of emerging technologies, the temporal tapestry of our digital age invites us to embrace curiosity, navigate complexity, and embark on a journey where the boundaries between the present and the future blur in the ever-accelerating march of technological progress.

Chapter 10

THE MYSTERIES THAT REMAIN

In the vast expanse of the temporal landscape, mysteries persist—questions that elude easy answers and beckon us to venture into the uncharted territories of understanding. In this chapter, we confront the enigmas that continue to captivate the curious minds of scientists and philosophers alike. From the fundamental nature of time to cutting-edge research and the open horizons of exploration, the journey into the mysteries of time unfolds.

The Enigma of Temporal Origins

One of the enduring mysteries that captivates the human imagination is the question of temporal origins. What initiated the unfolding of time itself? Was there a temporal beginning, or does time extend infinitely into the past?

As we delve into this profound inquiry, we find ourselves navigating the cosmic timeline. The journey takes us to the cosmic microwave background (CMB), a relic radiation that echoes the early moments of the universe. Studying the patterns within the CMB provides valuable insights into the conditions of the cosmos, but the mystery of what triggered the initiation of time remains.

Scientists, armed with observatories and advanced instruments, peer into the depths of space and time, seeking clues about the cosmic birth. The quest for temporal origins drives ongoing research, inviting us to ponder the very essence of existence.

Time and Ultimate Reality

Philosophers and physicists alike grapple with the question of whether time is a fundamental aspect of ultimate reality or merely a human construct. Does time exist independently of our perception and measurement, or is it a product of our cognitive processes and cultural frameworks?

The exploration of time's relationship with ultimate reality leads us into the realm of theoretical physics. Theories such as loop quantum gravity and string theory offer unique perspectives on the fabric of spacetime, suggesting that the foundational nature of time might be intricately woven into the structure of the cosmos.

As we stand at the crossroads of philosophy and physics, the mystery deepens. Our understanding of time's role in the cosmic narrative invites us to contemplate the profound interplay between our conceptualizations of time and the underlying fabric of reality.

Temporal Symmetry: The Arrow of Time

The arrow of time, representing the asymmetry between past and future, remains a puzzle that challenges our intuitions about the nature of temporal progression. Why does time exhibit a unidirectional flow, moving from past to future, while the fundamental laws of physics appear to be time-symmetric?

The second law of thermodynamics, which states that entropy tends to increase over time, is often cited as a key player in the asymmetry of time. Entropy, a measure of disorder, provides a directionality to processes—the natural tendency for systems to evolve toward states of higher disorder.

While the arrow of time is intimately linked with our everyday experiences, the underlying reasons for its emergence within the fundamental laws of physics remain an open question. Scientists explore the frontiers of theoretical physics, seeking to unravel the mysteries that govern the arrow of time and its intricate relationship with the cosmic order.

Quantum Time: The Dance of Particles

In the quantum realm, time takes on a different character, challenging our classical intuitions. The interplay between quantum mechanics and the nature of time introduces intriguing questions about the measurement of time intervals at the smallest scales.

Heisenberg's uncertainty principle, a cornerstone of quantum mechanics, posits a fundamental limit to the precision with which certain pairs of properties, such as position and momentum, can be simultaneously known. This uncertainty extends to the measurement of time intervals, introducing a quantum fuzziness at the heart of temporal quantification.

The exploration of quantum time leads us to consider the nature of "now" at the quantum level. Unlike classical time, where the concept of an absolute "now" seems straightforward, quantum time introduces a level of indeterminacy, challenging our notions of temporal simultaneity.

Time Travel and Temporal Possibilities

The allure of time travel, a staple in science fiction, sparks both fascination and skepticism. While the concept of time travel captivates the human imagination, the exploration of its feasibility and implications remains a topic of scientific inquiry.

Einstein's theory of relativity, formulated in the early 20th century, introduced the notion that time can be influenced by factors such as gravity and relative motion. Theoretical physicists explore the implications of these relativistic effects on the possibility of time travel.

As we navigate the theoretical waters of time travel, paradoxes such as the Grandfather Paradox and the Twin Paradox emerge, challenging our understanding of causality and the consequences of altering the past. While time travel remains a speculative realm, the exploration of its theoretical underpinnings fuels ongoing discussions about the nature of time and the boundaries of possibility.

The Philosophy of Time

In the realms of existential philosophy, time serves as a canvas for contemplation on the nature of human existence. Existentialist thinkers, including Jean-Paul Sartre and Albert Camus, delved into the temporal aspects of human life, exploring concepts such as bad faith and the absurdity of existence.

Sartre's concept of "bad faith" highlights the tendency of individuals to deny their freedom and responsibility by conforming to societal expectations and routines. This existential reflection invites us to consider the ways in which our choices and actions shape the narrative of our lives within the temporal framework.

Camus, in "The Myth of Sisyphus," confronts the existential condition of living in a seemingly indifferent universe. The myth of Sisyphus, condemned to roll a boulder uphill for eternity, becomes a metaphor for the human struggle to find meaning and purpose in the face of life's repetitive and sometimes absurd nature.

Time and the Aesthetics of Existence

The philosophy of time extends beyond metaphysical and existential inquiries, inviting us to explore the aesthetics of existence—the ways in which we appreciate and engage with the beauty, impermanence, and transience of life.

Friedrich Nietzsche, a philosopher with profound insights into the human condition, introduced the concept of the "eternal recurrence." This idea challenges us to live our lives as if we would have to relive them over and over again. Nietzsche's exploration of the aesthetics of existence prompts us to approach life with a heightened sense of appreciation, recognizing the ephemeral nature of moments and the potential for profound beauty in our experiences.

Time and the Unanswerable Questions

As we navigate the philosophical landscape of time, we encounter questions that defy easy answers—queries that invite us to ponder the very fabric of our existence. What is the nature of time's relationship with consciousness? How does our perception of time influence our sense of self?

The exploration of these unanswerable questions opens avenues for interdisciplinary dialogue, bringing together insights from philosophy, psychology, and neuroscience. The interplay between time and consciousness remains a frontier where the subjective experience of time intertwines with the objective measures provided by scientific inquiry.

The Future of Temporal Inquiry

In the grand tapestry of temporal exploration, the mysteries that remain beckon us toward the future. The ongoing journey of understanding time transcends disciplinary boundaries, encompassing the realms of physics, philosophy, psychology, and beyond.

Cutting-edge research, conducted in laboratories and observatories around the world, continues to push the boundaries of knowledge. Breakthroughs in quantum physics, advancements in neuroscience, and the exploration of the cosmos offer glimpses into the intricate interplay between time and the fundamental fabric of reality.

The quest for a unified theory that reconciles quantum mechanics and general relativity stands as a pinnacle in the pursuit of understanding time at its deepest levels. As researchers strive to unravel the mysteries of dark matter, dark energy, and the cosmic order, the very nature of time's role in the cosmic narrative comes into sharper focus.

A Call to Continue the Journey

As we stand at the nexus of the known and the unknown, the mysteries that remain serve as beacons guiding us toward further exploration. The quest to understand time, with its complexities and enigmas, invites each one of us to participate in the ongoing journey.

Whether through scientific inquiry, philosophical contemplation, or the aesthetics of existence, we are all stakeholders in the exploration of time's mysteries. The call to continue the journey resonates across disciplines and across time, inviting future generations to contribute their insights to the ever-expanding tapestry of temporal understanding.

In the spirit of curiosity and wonder, we embark on a collective endeavor to unravel the intricacies of time. The mysteries that remain, far from discouraging our pursuit, inspire us to delve deeper, to question assumptions, and to cultivate a sense of awe in the face of the unknown.

As we peer into the vastness of temporal inquiry, we find ourselves connected to the timeless pursuit of knowledge that transcends epochs and cultures. The mysteries of time, with their allure and complexity, propel us forward, urging us to continue the journey—a journey where the unfolding narrative of time intertwines with the human quest for understanding, meaning, and the eternal pursuit of knowledge.

REFERENCES

1. Sidney Perkowitz (2018). *Time Examined and Time Experienced – Physics World*. Physics World.

2. (2020). *A Debate over the Physics of Time*. Quanta Magazine.

3. A.S. Eddington (1948). *The Nature of the Physical World*. Cambridge University Press.

4. J.E. Boodin (1905). *The Concept of Time*. The Journal of Philosophy, Psychology and Scientific Methods, 2(14).

5. Pradyumn (2020). *The World of Particles*. Notionpress.

6. Pradyumn (2019). *Journey through the Dark Monster*. Notionpress.

7. H.G. Wells (2004). *The Time Machine*.

8. Albert Einstein, Robert W. Lawson - translator (2002). *Relativity: The Special and General Theory*. Methuen & Co Ltd.

9. Henri Bergson, F.L. Pogson - translator (1957). *Time and Free Will: An Essay on the Immediate Data of Consciousness*. Riverside Press.

10. Carlo Rovelli, Erica Segre - translator, Simon Carnell - translator (2018). *The Order of Time*. Riverhead Books.

11. Sean Carroll (2010). *From Eternity to Here: The Quest for the Ultimate Theory of Time*. Penguin.

12. Julian Barbour (2001). *The End of Time: The Next Revolution in Physics*. Oxford University Press.

13. Daniel C. Dennett (2017). *Consciousness Explained*. Little Brown.